SpringerBriefs in Political Science

More information about this series at http://www.springer.com/series/8871

Marco Delmastro

On the Measurement of Social Phenomena

A Methodological Approach

Springer

Marco Delmastro
Italian Communications Authority (AGCOM)
Rome, Italy

ISSN 2191-5466 ISSN 2191-5474 (electronic)
SpringerBriefs in Political Science
ISBN 978-3-030-77535-3 ISBN 978-3-030-77536-0 (eBook)
https://doi.org/10.1007/978-3-030-77536-0

This Springer imprint is published by the registered company Springer Nature Switzerland AG
The registered company address is: Gewerbestrasse 11, 6330 Cham, Switzerland

Writing books gives me a great sense of freedom, because you're not tied to the boundaries in which a song necessarily moves

Achille Lauro

For Lulù, Milli, and Cippa

Preface

This book is written during the period of isolation that forced us to stay at home and to reflect on various aspects of our lives. I have reasoned on some elements that have characterised my scientific research and work over more than twenty years.

It is perhaps for this reason that I stood to reflect on some methodological and ethical issues of applied science. These days it is so palpable to see how important the data are, how their analysis is key to orientating our daily actions, and that, often, behind each of them, there is an event, a story, there are feelings, that we have the duty to analyse with care and diligence.

I chose the threats to journalists as one the "fil rouge" of my reasoning because it represents a crucial social issue, especially in Italy, where some events, such as the murder of journalist Giancarlo Siani, have marked an entire city (Naples) and, more generally, an entire country.

Public and private violence episodes are difficult to study rigorously, and therefore represent the best test case for a digression of applied methodology. On the one hand, there is the temptation to put too much in the analysis, to the point of distorting the results to corroborate our ideological beliefs (or prejudices). This is known as "research bias", i.e., a process where the scientists performing the research influence the results, to portray a certain outcome. Of course, this problem is very much present in science, especially in the social sciences, where issues with an ethical dimension are very often addressed. In addition, applied sciences are undergoing a period of great ferment and revolution in methods and tools. This is due to many factors, in large part to the emergence of the use of the so-called "big data": an unprecedented volume of data and information from different sources and in various formats now provide insights into many individual and social dimensions that previously could not be examined. In this case, the research bias is often embodied in a widespread reluctance to change, which has always distinguished scientific evolution, according to a leapfrog pattern described by Thomas Kuhn.

On the other hand, social scientists run the opposite risk of putting too little, because if they focus only on technical issues, they will inevitably move away from the phenomenon under study, for mere intellectual speculation, or even worse, for a few more scientific publications.

Over the years, I believe I have been the victim of both "temptations", and I fear I am in good company, given the rising lack of trust in scientific research and the paradoxical proliferation, in an era of scientific drive, of conspiracy and unscientific theories.

Research implies a very difficult balance between care of data, respect for the stories behind them and ideological and research biases which, even in absolute good faith, direct the analyses.

The use of non-robust contagion data today seems to me to be disgraceful. Uncertain classifications, inaccurate measurement systems, and yet this piece of information is provided and disseminated every day throughout the world. Important individual and collective decisions are made upon them, and political, national and international, disputes arise.

My feeling is that we are in too much of a hurry to find solutions, especially solutions that reflect our prior beliefs, so that we remain in the riverbed of what we already know. But often life surprises us with new situations, different points of view and elements that we cannot fully understand.

To admit it would be a first step to sweep away conspiracy explanations; to open ourselves to different points of view, a stimulus to build more solid and socially shared scientific explanations.

The reflections that follow do not contain revolutionary points of view or new theories, they are wide-ranging considerations aimed at identifying a "modus operandi" that could be useful to deal rigorously with the analysis of social phenomena.

Enjoy the reading!

Rome, Italy Marco Delmastro

Acknowledgements

I am grateful to several people for their help with this book and more generally for discussing with me general issues of applied science.

First, I am indebted to all colleagues of the Department of Economics and Statistics of Agcom. In particular, Otello Ardovino, Emanuele Brugnoli, Alessia Leonardi and Gennaro Ragucci have improved with their corrections and suggestions this book beyond measure. Similarly, suggestions from Fabiana Zollo were invaluable. I also want to thank Jacopo Arpetti, Elisa Colorito, Stefano D'Alfonso, Raffaella Della Morte, Donatella Di Lieto, Ilaria Fevola, Aldo Fontanarosa, Adriana Lotti, Francesco Marrazzo, Sara Mastrovita, Rocco Mazza, Giuseppe Mennella, Gianmarco Pondrano, Attilio Scaglione, and Alberto Spampinato, for suggestions and support on some contents of the book.

For the part concerning the analysis of threats to journalists, I thank the following institutions: FNSI (Federazione Nazionale della Stampa), INPGI (Istituto Nazionale di Previdenza dei Giornalisti Italiani), ISTAT (Istituto Nazionale di Statistica), Ordine dei Giornalisti, Ossigeno per l'Informazione, USGF (Unione Sindacale Giornalisti Freelance) and USIGRAI (Unione Sindacale Giornalisti Rai). Special thanks are due to Mimma Iorio and Federico Polidoro.

The book also benefited from discussions on specific topics with Stefano Castriota, Massimo G. Colombo, Mario Morcellini, Domenico Piccolo, Antonio Scala, Rosaria Simone, Sergio Splendore, Mirco Tonin, and Giorgia Zamariola.

I am very grateful to Niko Chtouris and the team of SpringerNature for their patience and support.

This book is dedicated to Keith Cowling, for sharing with me his passion, ethics, and knowledge of social sciences many years ago when I was just a young Ph.D. student, lost before science.

Contents

About the Author

Marco Delmastro is the Managing Director of the Department of Economics & Statistics at AGCOM (the Italian regulator for the communications services). Before, he worked in Agcom as the President's Economic Advisor. Before joining Agcom, he was economist at the Italian Competition Authority (AGCM), and case handler of antitrust cases in the communications sector. He spent few years at Politecnico di Milano as a Senior Research Fellow and at the University of Pavia as a Research Fellow.

He received both a Master degree in Economics at Collegio Carlo Alberto (Turin) and a Doctoral degree in Economics at the University of Pavia. He also holds a Ph.D. in Economics at the University of Warwick (UK).

He is author of numerous scientific publications on topics that cover industrial economics, competition and regulation, sociology, computational psychology and behavioural sciences.

With Professor Massimo Colombo he is co-author of "The Economics of Organizational Design", published by Palgrave Macmillan. More recently, he has published (joint with Prof. Antonio Nicita) a book on "Big Data", edited by Il Mulino.

For these research achievements, he has obtained national scientific qualifications as a Full Professor in Economics, Political Economy, and Applied Economics.

List of Figures

List of Tables

List of Boxes

Chapter 1
Introduction

We live in a world crowded with data and numbers. Every day we make decisions based on data and measures that encapsulate them. We take our temperature, we calculate the distance from home to a place we want to go, we check the statistics of our favorite sport teams, we read the dosage of the ingredients of a recipe. Likewise, governments make decisions based on indicators such as the gross domestic product, the unemployment rate, investments in research and development, the poverty index.

In recent years, a new massive wave of big data from online human behavior (i.e., the digital footprint) is being used by companies, researchers and institutions for a wide variety of purposes, some good, some not so good. This data is employed to generate metrics that are used to cluster individuals into categories, predict their interests and anticipate behaviors, assess social interactions, even forecast weather conditions or provide estimates of influenza activity.

In addition, the outbreak of COVID-19 has made essential indicators such as the number of confirmed cases, death and fatality rate, even the "Rt value", based on which governments around the world have taken health, social and economic policies, and citizens have been forced to make many sacrifices, even at the cost of having some fundamental freedoms severely restricted.

In short, public policies, daily choices of citizens, the strategies of companies and institutions are increasingly "data-driven".

In this context, knowing the methods that are used to compute the measures that shape our lives is not only of interest to researchers but also of fundamental importance to citizens and society as a whole.

In this book, we show how the operational classification of concepts like unemployment or poverty is susceptible to focusing public policies on some aspects while neglecting others, with profound effects on economic and social actions supporting different social categories. For instance, the emphasis on the gross domestic product as a measure of a country's wealth conceals a precise political approach which, although it has been frequently disputed by some researchers, has never been the subject of a real public debate.

M. Delmastro, *On the Measurement of Social Phenomena*,
SpringerBriefs in Political Science,
https://doi.org/10.1007/978-3-030-77536-0_1

The same applies to many commercial policies of companies when they assume a personal and public value. In recent years, the burgeoning economic and social role of online platforms (such as search engines, social networks, and e-commerce websites) has led to demand for greater transparency about the algorithms and metrics that govern aspects such as user profiling, personalization of services, and recommendations of searches and news information.

Methodologies for measuring social phenomena is not just a problem for experts. Understanding the rationale and limits of measurement, discussing the most appropriate indicators according to the objectives that society sets for itself, are all issues that are too often neglected, because deemed to be technical, but which are of fundamental value for our democracies.

Greater discussion and transparency on the indicators governing the policies of governments and companies, on their operational classifications, on how they are measured and then used, requires a holistic approach to the theory of measurement of social phenomena. Unfortunately, this theory is dispersed among many scientific fields and lacks overall coordination and overview.

Philosophy discusses the general criteria of several fundamental aspects of measurement, ranging from the classification of phenomena, the feasibility of measuring them, up to the methodological approach itself. Furthermore, the relationship between different phenomena (i.e., association versus causation) is a central aspect of the philosophy of science. Statistics offers many of the tools for the effective measurement of phenomena and the estimation of relationships between them, ranging from sampling theory to inferential statistics and econometrics. In the last few decades, computer science has also become essential for measuring social phenomena because it enables the gathering, processing and analysis of huge volumes of data from various sources (e.g., social networks, e-commerce websites, geolocation) and of various formats (e.g., text, video, images) concerning everyday individual and social life, by using methodologies ranging from network science models to those of natural language processing. Finally, all sciences have developed specific procedures for measuring indicators in their various fields of interest: in social sciences, in fields such as economics (e.g., unemployment rate, poverty index, income inequality), psychology (e.g., psychometric indicators), and sociology (e.g., life satisfaction, crime rate).

This impetuous evolution of measures and indicators, which has been further strengthened in recent years by the increasing availability of data, has developed in the scientific community within fragmented and sometimes not sufficiently sound approaches.

The aim of the book is therefore to shed light on some essential methodological steps. The book builds up in such a way as to provide the reader with an integrated and original interpretative compass, hinting at the most important technical aspects and referring to the relevant literature for further in-depth study.

In Chap. 2 we discuss a three-step procedure for measuring social phenomena. The procedure consists of three consecutive steps that address: (i) the definition and classification of social phenomena, which are often vague and abstract concepts in need of operationalization; (ii) the definition of measures of phenomena as defined in the first step; (iii) the actual measurement procedures. The various steps must be discussed not only to assess their scientific robustness but also to understand the perimeter and limits of the final measurement of the underlying concepts. For example, it is widespread in the social sciences the existence of a shadow part of social phenomena, even very significant, which is often not recorded by official statistics (e.g., the shadow economy). Any scientific reflection and political strategy based on indicators must take due account of how much the available indicators explain the overall phenomenon, and how much is not covered by them. This is only possible when the various steps of the measurement procedure have been duly addressed and clarified.

Social phenomena cannot be analyzed in isolation but are multifaceted and multi-dimensional. In Chap. 3, therefore, we first address the issues related to the selection of dimensions to be taken into account in the analysis, and the identification of procedures that can be used to analyze multifaceted phenomena using composite indicators. Once these steps have been clarified, we come to the fundamental question of what kind of relationship exists between the different dimensions. This is not only a statistical question but also a more general philosophical dilemma that is rooted in Descartes, and rationalism, on the one hand, and in Hume, and empiricism, on the other. Different scientific approaches (e.g., rationalism versus empiricism) have strongly shaped the evolution of knowledge in fundamental fields of social science such as decision theory. In the case of relations between phenomena, the debate spans from simple associations to cause-effect (i.e., causation) relations.

The emergence of big data analysis has not only provided the social scientist with a multitude of data on (large parts of) the population and a wide range of domains but has also imposed associative relationships. It is both a statistical problem of estimating the relationships between a huge number of variables, and the recognition that the social system is complex and it hardly fits into a simple causal model in which a few variables are considered in isolation.

This observation leads to a further aspect that is one of the leitmotifs of the book: the presence of research, and consequently, policy biases. For a long time, decision theory avoided dealing with the evidence (the first and maybe still most famous being Allais' paradox in 1953) which showed that the underlying rationalist axiomatic structure did not fit with the way individuals make their own decisions. Decades of cognitive theory in psychology and neuroscience have confirmed this fact, which was already evident at the time of Hume. The discussion has shifted over the years from descriptive to normative aspects, trying to use the rationalist approach in terms of optimal decision-making. However, cognitive science has shown that the non-rationality of human behavior must also be taken into account from a normative point of view (see, for instance, the nudge theory).

This is to say that modern science needs what the Italian economist and sociologist Vilfredo Pareto, famous for his theory of welfare, recognized at the beginning of the twentieth century as the future of decision theory, namely the integration of knowledge (in his case the foundation of economics by psychology).

In Chap. 4, we, therefore, address the problem of integration in the measurement of social phenomena. Data integration has been an essential part of applied science for some years now. Even though in many scientific fields there is still a considerable amount of reluctance due to the inertial evolution of knowledge and the challenge of leaving established tools and embracing new ones, a growing part of applied research combines traditional methodologies and data with new tools and information. However, this process will only be achieved when knowledge is fully integrated, and the current compartmentalization of knowledge and scientific organization is replaced by a new one. The specialization of knowledge has brought great scientific benefits. However, the time has come to define new models that combine specialization, which remains necessary, with forms of increasing and comprehensive integration.

Chapter 2
A Three-Step Procedure

Social research is often focused on complex technical aspects. However, the applied study of social phenomena raises several preliminary methodological issues which, if not effectively addressed, risk undermining the quality of the whole analysis. These questions, on the contrary, are rarely addressed by researchers and are generally taken for granted, referring at most to the relevant scientific literature. Since, each study tends to address these issues differently even within the same line of research, reference to the relevant literature is insufficient if not misleading.

In this chapter we provide illustrative examples to show that failure to address these methodological steps is susceptible to produce severe research pitfalls (see for example cases illustrated in the boxes). Moreover, we use a rather intricate social phenomenon (i.e., threats to journalists) to exemplify how these issues can be addressed and how different approaches lead to very different results.

The overall recommendation is that any applied analysis of social phenomena should contain, in the beginning, a clearcut outline of how these issues are addressed in the research.

Briefly, this preliminary methodological process, that we call here the "three-step procedure", can be summarized in the following, successive and concatenated three operational steps.[1]

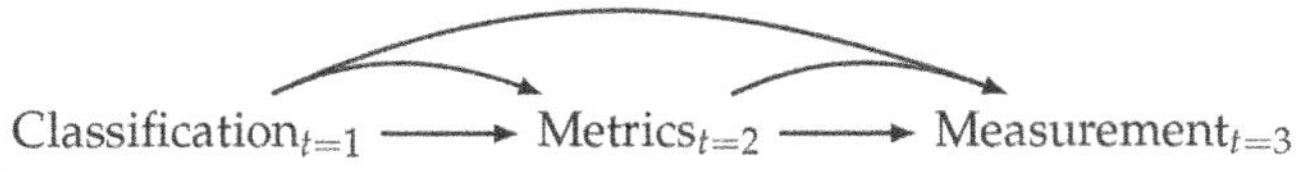

Firstly, inaccurate or imprecise definitions and classifications of social phenomena risk either excluding important aspects from the analysis or including misleading elements. Consequently, the exact definition of the perimeter under analysis is crucial

[1] Sometimes these steps are called: (i) characterization (or ontology); (ii) representation; and (iii) procedures. See Cartwright and Bradburn (2016) and National Research Council (2011).

M. Delmastro, *On the Measurement of Social Phenomena*,
SpringerBriefs in Political Science,
https://doi.org/10.1007/978-3-030-77536-0_2

for the understanding of the phenomena under observation (see Sect. 2.1). Moreover, the definition must necessarily be both operational and dynamic, i.e., it should contain the operational classification of the phenomenon and it should possibly change over time, according to the evolution of the economic, normative, social, and/or technological factors.

Once the phenomenon has been classified, it must be correctly measured (Sect. 2.2). This means that a system of metrics through which quantitatively evaluate the different dimensions must be defined in advance. In this sense, metrics must give rise to numerical values that can be bivariate (0/1, in the case of TRUE/FALSE logical values), natural, integers, or real numbers, or that can be ordered according to a categorical scale (for instance, on an x-item scale to evaluate the frequency of an event: e.g., "never", "sometimes", "once a week", "several times a week", "every day").

After the identification of the metric system, the actual measurement should be carried out through an appropriate measurement methodology (Sect. 2.3).[2] This stage, that is usually the one discussed in the "method section" of a scientific paper, should tackle multiple problems that may arise in the measurement process. In social science, these are mainly statistical, but they may also involve more general problem of analysis, such as the presence of a shadow component of the phenomenon under investigation (Sect. 2.3.1).

In this chapter, we discuss the three methodological steps in a rather informal and non-technical way, suggesting more technical references, providing illustrative examples of important social phenomena (presented in the boxes) and taking the threats to journalists as a case study that serves as a *fil rouge* throughout the chapter (and the book).

2.1 Classification of Social Phenomena

> Classification is the essential first step in science. (Szostak 2004)

Special attention should be paid to the initial process of definition and operational classification of the phenomenon, which is often much more complex than it may seem at first sight. This step directs all the subsequent analyses and must therefore follow an accurate methodology that should be clearly presented to the reader.[3]

Sometimes classifications may be taken as self-evident. However, important cases, such as those summarized in the following boxes, show that most of the time they are not. How exactly do we define the disinformation phenomenon? What do we mean by Covid-19 cases?

---

[2] It is worth noticing that measurement is often associated with both the definition of the metric system and the procedures aimed at measuring aspects of the empirical world.

[3] "*The history of science teaches the importance of classification, both as a phase of individual mastery, and as an aid to the further progress of comprehension*" (Hayes, E. 1911).

In social science, many phenomena (such as poverty, race, social exclusion) can be described by concepts that are fuzzy and multifaceted and for which the boundaries are not clear (i.e., "Ballung concepts"): "*the use of concepts for different purposes, often leads to changes in characterization, representation and/or procedures that disrupt the possibility of common metrics*".[4]

This is often true for the most widely analysed and discussed concepts in economic and social literature.

Take the case of unemployment. When we consider unemployment, we are thinking of a very broad and multifaceted concept. However, its operational definition is "*the unemployed comprise all persons above a specified age who during the reference period were: without work, that is, were not in paid employment or self-employment during the reference period; currently available for work, that is, were available for paid employment or self-employment during the reference period; and seeking work, that is, had taken specific steps in a specified recent period to seek paid employment or self-employment*" (OECD, Glossary of Statistical Terms). Note that this definition, for example, leaves out all persons who have stopped actively looking for a job. This limitation could be very relevant, especially in contexts of endemic labour shortages.

Therefore, by moving from the broad theoretical concept of unemployment to its operational definition, we would limit the analysis to specific conditions. This methodological choice inevitably directs the result and must therefore be discussed in advance in all its implications.

To make this definition an operational classification we need to identify the age beyond which we consider individuals (most of the time this age is set at 15) and the reference period. This additional step further specifies and delimits the concept of unemployment to be measured.

So, we have started from a very broad concept of unemployment and we have ended up with a measurement of a specific part of it. When we associate the concept with its operational metric, we must remember how it was defined and classified in order to identify possible fallacies of the whole subsequent measurement process.[5] This is particularly important because unemployment, like many other social phenomena (e.g., misinformation, the spread of Covid-19), is a social phenomenon which receives great public attention and upon whose metrics important political decisions are made.

---

[4] See Cartwright and Runhardt (2014). The term Ballung concept is borrowed by the authors from Otto Neurath; see Jordi, C. (2021).

[5] "*These should be the best of times for working people. After all, the unemployment rate is just 3.7 percent, near its lowest level in almost 50 years. If that were our only guide, then we might conclude that virtually every adult in the United States who wants to work is either doing so or is diligently looking for a job. The problem is that the unemployment rate, which is publicly updated by the Labor Department on the first Friday of every month, is an insufficient statistic, despite all the attention it gets. Sure, it tabulates the number of men and women who are actively job-hunting. But it leaves out the growing numbers who have stopped actively looking*" (Louis Uchitelle, Unemployment Is Low, but That's Only Part of the Story, *New York Times*, July 11, 2019).

In Box 2.1, we show how different national classifications of Covid-19 cases and deaths may be one of the reasons behind the observed geographical variability.

Box 2.1 Issues in the classification of Covid-19 cases

The World Health Organization defines Covid-19 cases and deaths as follows *Confirmed case*: A person with laboratory confirmation of COVID-19 infection, irrespective of clinical signs and symptoms; *Death due to COVID-19*: a death resulting from a clinically compatible illness in a probable or confirmed COVID-19 case, unless there is a clear alternative cause of death that cannot be related to COVID-19 disease (e.g., trauma). There should be no period of complete recovery between the illness and death (WHO, Public health surveillance for COVID-19).

As shown in the figure below, in March 2020 there were significant differences between countries in the death rate from COVID-19. These differences have been the subject of lively debate, with potential drivers such as age distribution, history of previous illnesses and the level of smog in the atmosphere being highlighted. Obviously, these factors may have significantly influenced COVID-19 death rates, but there are also differences in the operational classifications adopted by each country. These differences are due to the fact that the WHO's definition inevitably leaves a certain degree of discretion in the actual process of classifying the phenomenon. The latter should be considered before benchmarking data.

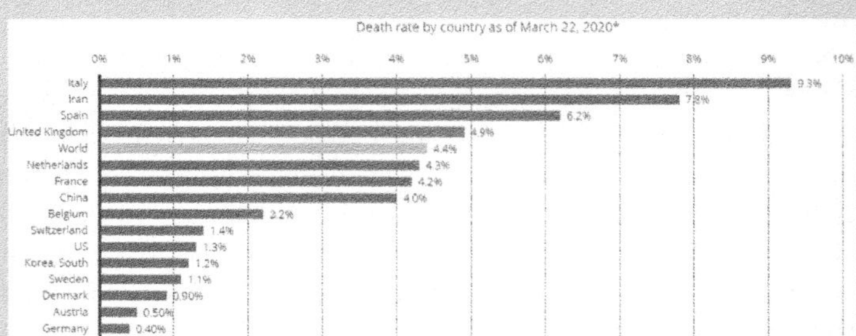

Source STATISTA

As the figure shows, Italy, in March 2020, was by far the country with the highest death rate. However, Prof. Walter Ricciardi, the Italian representative to the WHO, mentioned in an interview that «*The way in which we code deaths in our country is very generous in the sense that all the people who die in hospitals with the coronavirus are deemed to be dying of the coronavirus*». Krys Johnson, epidemiologist at Temple University College of Public Health, suspects if different classifications and measurement systems were taken into account «*the true fatality rate in Italy is closer to the global fatality rate of 3.4*» (see Rachael Rettner, "Why Deaths from Coronavirus Are So High in Italy", *Scientific American*, March 10, 2020).

The representation of a social phenomenon, through its definition and classification, must be appropriate to the central features taken to characterize it. It is important to ensure that the final items measure the concept originally intended. This step is both a process and an end result.

Here are a few basic rules of the classification process. Bailey (1994) can profitably be followed (see also Szostak 2004).

First, classification may be defined as the ordering of entities into groups or classes based on their similarity. The aim is to minimize within-group variance, while maximizing between-group variance. In other words, when we arrange a set of entities of a phenomenon into categories, we want that each group is as different as possible from all other categories, but each category is internally as homogeneous as possible. This process can be done following a theoretical ordering (based on knowledge of the phenomenon), or statistically (e.g., through cluster analysis). In any case, "*one basic secret to successful classification, then, is the ability to ascertain the key or fundamental characteristics on which the classification is to be based [...] In all of these diverse cases, prior knowledge and theoretical guidance are required in order to make the right decisions*" (Bailey, 1994).

Second, the defined categories must be both exhaustive and mutually exclusive. This means that if *N* events have to be classified (e.g., type of threats to journalists, see Sect. 2.1.1), there must be an appropriate category (i.e., exhaustivity) and only one for each, with no event being in two different categories (i.e., mutual exclusivity). In other words, the classification of the *N* events into categories must aim to cover all the events, and each event must be included in one and only category. However, as we have seen in the case of unemployment, this is not always the case. In many cases, classification is not exhaustive, and this implies that there is a part of the phenomenon that is not detected by the subsequent measurement step. This is the case of important parts of social phenomena that remain hidden from official statistics (see par. 2.4).

Nowadays, automatic clustering is widespread (e.g., K-Means clustering, hierarchical clustering),[6] especially in data mining. However, this process, if not governed by an appropriate knowledge of the phenomenon, may lead to inconsistent (as well as ethically questionable) results. Moreover, data clustering does not imply classification of social phenomena, and this aspect is often absent in data mining analyses, to the extent that the "end of the theory" has been suggested.[7]

---

[6] See Allahyari et al. (2017).

[7] See Anderson, C. (2008). The End of Theory: The Data Deluge Makes the Scientific Method Obsolete, *Wired Magazine*, 16. For a discussion, see, among others, Mazzocchi, F. (2015). Could Big Data Be the End of Theory in Science? *EMBO Reports*, 16, 1250–1255, and Coveny, P., et al.

Box 2.2 Issues in the classification of mis- and disinformation

Information disorder is one of the hot topics of social sciences: from media economics to sociology of communication, from politics to data science, from cognitive psychology to political economy. In recent years, there has been an astonishing number of articles on the subject that have been published in the major journals in these fields (Science, Nature, PNAS, Journal of Economic Perspectives, Cognition, etc.). However, as brilliantly argued by Wardle and Derakhshan (2017) «*One depressing aspect of the past few months is that, while it has resulted in an astonishing number of reports, books, conferences and events, it has produced little other than funding opportunities for research and the development of tools. One key reason for this stagnation, we argue, is an absence of definitional rigour, which has resulted in a failure to recognize the diversity of mis- and disinformation, whether of form, motivation or dissemination*».[8]

I would add that it is not only a problem of definitional rigor, epitomized by the term "fake news" that is used indistinctly to indicate a wide range of concepts,[9] but also of operational classification once this phenomenon has to be analyzed empirically.

In this respect, disinformation starts from a process of ex-post verification[10] of factual information by independent (third party) and trusted fact checkers/debunkers. It can be either "objective" (i.e., directed at verifying single news items) or "subjective" (i.e., directed at verifying news sources).

Clearly, the analysis of information disorder substantially differs depending on the choice of the operational classification.

(2016). Big Data Need Big Theory Too, *Philosophical Transactions of the Royal Society* A, 374: 20160153.

[8] "Misinformation" is when false information is shared, but no harm is meant; "disinformation" is when false information is knowingly shared to cause harm; "malinformation" is when genuine information is shared to cause harm, often by moving information designed to stay private into the public sphere. See, Wardle & Derakhshan (2017). Information Disorder: Toward an Interdisciplinary Framework for Research and Policy Making, *Council of Europe Report*, DGI 09, p. 5.

[9] News coming from sources which have not been professionally and adequately verified; satirical news that, out of context, may be perceived as real by citizens; news coming from sources fueling conspiracy or plotting theories; news coming from sources specialized in gossips or rumors and pseudoscience, characterized by unverified current events or scientific breakthroughs; hate news coming from sources promoting racism, misogyny, homophobia and other forms of discrimination; (entirely or partially) correct news with sensationalist titling for click-baiting purposes; news coming from sources providing biased information to support determined opinions and political orientations.

[10] Internal fact-checking instead is verification done in-house by the publisher.

2.1.1 Classification of Threats to Journalists

In this essay, we present a case study to indicate how some methodological concepts can be operationalized. We have chosen a social phenomenon that embodies many of the major issues that social science practitioners may experience in their research work.

Harassment and intimidation of journalists is indeed a social phenomenon in which: (i) there are no official statistics (if not very incomplete); (ii) international data are difficult to compare (except for the most violent acts such as killings); (iii) the perimeter of the universe of reference needs to be defined; (iv) there is a very significant hidden/shadow component that could mislead the analysis itself (see Sects. 2.3.1 and 2.4.1). In addition, the phenomenon has far-reaching social implications that go well beyond its realm and affect the overall news production and the fight against pathological social phenomena such as organized crime and administrative and political corruption (see Chaps. 3 and 4).

While the definition of the phenomenon is quite straightforward (i.e., any act or action aimed at intimidating journalists in their profession) the operational classification is more complex.

In this respect, we develop a classification based on national and international NGOs monitoring systems (e.g., Ossigeno per l'Informazione, Reporters Sans Frontier) which have been investigating intimidations of journalists for years. This categorization has been adopted at international level and used in monitoring projects under the aegis of the European Commission and the Council of Europe.[11]

The classification has been then integrated with some cases identified at international level,[12] with particular reference to online intimidation,[13] in order to detect also new forms of threats through the web, thus having a picture of the phenomenon as exhaustive and updated as possible.

In addition, the classification was validated by journalists and experts in the field,[14] shared with international institutions (i.e., UNESCO) and discussed at international meetings.[15]

A schematic illustration of the operational classification of threats to journalists is summarized in Table 2.1.

The classification distinguishes two basic categories (i.e., direct threats and indirect threats) and four second-level categories[16]: (1) traditional threats (i.e., killings,

---

[11] Council of Europe (2019). *Democracy at Risk: Threats and Attacks Against Media Freedom in Europe*, Annual Report by the Partner Organisations to the Council of Europe Platform to Promote the Protection of Journalism and Safety of Journalists.

[12] Reporters Sans Frontieres (2019). *World Press Freedom Index 2019: A Cycle of Fear*, 2019.

[13] ECPMF (European Centre for Press and Media Freedom) (2020). *Mapping Media Freedom.*

[14] Raffaella Della Morte, Ilaria Fevola, Aldo Fontanarosa, Adriana Lotti, Francesco Marrazzo, Giuseppe Mennella, Alberto Spampinato, Sergio Splendore are gratefully acknowledged.

[15] See World Press Freedom Day, Rome, May 8, 2018.

[16] *Risks and threats to journalists themselves come in many forms*; Wahl-Jorgensen, et al. (2016). The Future of Journalism, *Journalism Studies*, 17, 801–807.

Table 2.1 Classification of threats against journalists

Categories	Second level	Third level
Direct threats	Traditional	Aggression (killings, violent actions, shots, ...)
		Warning (verbal warning, threatening letter, death threats, ...)
		Damage to personal property
	Web	Diffusion of false personal information
		Offence
		Sexual offence
		Threats of violence (including sexual violence)
Indirect threats	Legal actions	Reckless lawsuit
	Obstacle to information	Impeding access to documents and data

warnings, assaults, damage to personal property); (2) intimidation on the internet (i.e., web offences and threats, digital diffusion of false information); (3) legal actions (i.e., reckless complaints); (4) obstacles to the acquisition of information and documents. The first two can be referred to as "direct threats", while the latter can be called "indirect threats".

The full range of these actions defines the whole perimeter of the types of threats to journalists (exhaustivity), including physical and virtual intimidation, concrete actions and legal acts, direct acts and indirect obstacles. Moreover, the classification fulfils the criterion of exclusivity since each event is assigned to only one category.

It should also be noted that some analyses carried out on a worldwide basis report the number of journalists killed (or any direct action intended to endanger the physical safety of journalists).[17] In these latter cases, the classification of threats is limited only to the extreme cases, leaving aside all other forms of threats (i.e., it does not meet the criterion of exhaustiveness). It is self-evident that such a methodological approach, even if necessary in some countries, due to the infeasibility of a different, and finer, monitoring system, is not much informative, excessively delimiting the perimeter of the phenomenon under study (to a small subset of direct threats), and limiting, in the last instance, the scope of the analysis, both from a dynamic and comparative points of view.

2.2 Metrics of Social Phenomena

The increasing availability of individual data has livened up the debate on metrics in recent years (for a discussion see, for example, Bohrnstedt 2010 and National Research Council 2011). The emergence of big data has led to what has been called a process of "datafication", i.e., "*the process by which subjects, objects, and practices*

[17] See, for example, INSI (International News Safety Institute) (2019). *Killing the Messenger.*

are transformed into digital data" (Schintler et al. 2021). This process has amplified the need to create new methods for robust social metrics (so called smart metrology).

To this aim, the set of traditional methods used for measuring social phenomena has been extended by the increasing employment of metrics from heterogeneous fields, such as cognitive psychology (see Box 1.3), neuroscience[18] and data science[19] (see Van Bavel et al. 2020, for a methodological proposal of a broad behavioral approach to study the reaction of society to Covid-19 epidemic and related policies).

Box 2.3 Psychometrics: the mood and feelings questionnaire (MFQ)

Measurability in psychology traces back at least to Gustav Fechner (1860) (in his *Elements of Psychophysics*), in which he illustrated methods of measuring intensities of feelings. Since then, psychophysicists have been working extensively on methods for measuring feelings and mood. One of the most relevant is the Mood and Feeling Questionnaire (in two forms: long and short)[20] which includes 33 (13 in the short version) items indicating how much individuals have felt mentally distressed during the previous two weeks (e.g., "I felt miserable or unhappy", "I didn't enjoy anything at all", "I found it hard to think properly or concentrate"). The items include general statements related to mood and feelings and might indicate the presence of clinically relevant disorders such as depression and anxiety. The answers are given on a three-point Likert scale where respondents are asked to decide if the statements are "true", "sometimes true", or "not true". The measurement scale is thus ordinal as in most cases in psychometrics.

Scoring of the MFQ is obtained by summing together the point values of responses for each item (i.e., additivity). The response choices and their designated point values are as follows: "not true" = 0 points, "sometimes true" = 1 point, "true" = 2 points.

---

[18] In cognitive neuroscience, metrics such as the firing rate of the neuron (i.e., the number of spikes per time) and related graphical and statistical methods (e.g., a Raster plots, a histogram which shows action potentials as a function of time) are widespread (see, M. Gazzaniga et al. (2018). *Cognitive Neuroscience: The Biology of the Mind*, Norton).

[19] Social science has recently been revolutionized and revitalized by the use of data science applied to social phenomena; in this vein, see Cappella (2017) for communication studies; Cox et al. (2018) for statistics, Edelmann et. al (2020) for computational social science; Halford and Savage (2017) for sociology, Varian (2014) for economics.

[20] See Angold, A., Costello, J., Van Kämmen, W., & Stouthamer-Loeber, M. (1996). Development of a Short Questionnaire for Use in Epidemiological Studies of Depression in Children and Adolescents: Factor Composition and Structure Across Development, *International Journal of Methods in Psychiatric Research*, 5, 251–262.

Clinical studies have shown that higher scores on the MFQ indicate more severe depressive symptoms (i.e., validation[21]). The range of scores on the MFQ varies from 0 to 66 (26). A score of 27 (12) or higher may indicate the presence of clinically relevant depressive symptoms in the respondent.[22] MFQ has been also recently used to measure the emotional impact of COVID-19 on the population.[23] The MFQ has a number of important (psychometric and implementation) features: internal consistency, test–retest reliability, validity, sensitivity to change.

It is therefore impossible to review in a few pages (and even probably in a whole book) the growing range of literature on measurement. Here, it is important to sum up some background considerations.[24]

First, since social science concepts are large in number and fuzzy (see above "Ballung concepts") "*researchers have not yet discovered how to define the kind of fundamental quantities in the social sciences that exist in the physical sciences*" (National Research Council 2011, Ch. 2). Indeed, compared to physical metrics, social metrics generally present a lower degree of objectivity and standardization and are characterized by a higher variety. They are based on a large spectrum of methodologies and techniques and include socio-demographic attributes (e.g., age, gender, marital status), economic characteristics (e.g., income, wealth, unemployment, poverty), up to psychological traits (e.g., intelligence, mood, depression).[25]

This realm has been investigated deeply by the so-called "measurement theory" (or mathematical theories of measurement) that in particular explores the study of measurement scales.

Numerical scales can be of various types: (i) nominal (i.e., in which the dimensions of a phenomenon are assigned to classes without a specific order; e.g., gender); (ii) ordinal (i.e., dimensions are ordered but with no underlying algebraic structure; e.g., education); (iii) interval (i.e., equal intervals between objects represent equal differences, but not ratios; e.g., temperature) and (iv) ratio (i.e., ratio scales where zero

---

[21] Turner, N., Joinson, C., Peters, T. J., Wiles, N., & Lewis, G. (2014). Validity of the Short Mood and Feelings Questionnaire in Late Adolescence, *Psychological Assessment*, 26, 752–762.

[22] Thabrew, H., Stasiak, K., Bavin, L. M., Frampton, C., & Merry, S. (2018). Validation of the Mood and Feelings Questionnaire (MFQ) and Short Mood and Feelings Questionnaire (SMFQ) in New Zealand Help-Seeking Adolescents, *International Journal of Methods in Psychiatric Research*, 27, 1–9.

[23] Delmastro, M., & Zamariola, G. (2020). Depressive Symptoms in Response to COVID-19 and Lockdown: A Cross-Sectional Study on the Italian Population, *Scientific Reports*, 10, 22457.

[24] The study of measurement is massive and embraces scientific domains ranging from the epistemology of science to data science (it also involves statistics, mathematics and all the hard and social sciences in whose field the various measures are developed). Tal (2020) presents a pleasant and accurate review from which the interested reader may start. See also Mari (2005) for issues related to the epistemological foundation of measurement.

[25] These aspects (in particular, the variety of concepts and methodologies employed in social sciences) are further analyzed in the next chapter where composite indicators will be developed.

is meaningful; when they have a natural unit, such as those representing probabilities; they are called absolute scales).[26]

In this regard, the development of measures can be seen as a process of assigning numbers to represent the qualities of the phenomenon under investigation (Campbell 1957).

One lesson we can learn is that empirically significant aspects of a measurement scale are those that mirror relevant relations among the events of a phenomenon being measured. The issue is how to express the events that characterize a phenomenon into a numerical scale, i.e., how to create a mapping that associates real world events with numerical structures, so that this association respects the structure of the events that make up the phenomenon and at the same time the numerical scale is characterized by nice properties (such as additivity).

In this context, any measurement process that satisfies the conditions of additivity is called "fundamental". To do this, ordering and concatenation are sufficient for the construction of an additive measure of a dimension of the phenomenon (e.g., length, duration, weight, electrical resistance). Other measurement procedures can be indirect, because derived from relations to other fundamentally measurable dimensions (e.g., temperature, density).

However, this cannot be the case for a vast array of social phenomena, in which only ordering is feasible, so that additivity cannot be fully achieved.

The debate on measurement procedures has thus developed, especially by virtue of the development and diffusion of psychometric metrics (see Box 2.3), to the point of including procedures based on ordinal scales (i.e., which do not satisfy, even indirectly, the criteria of additivity).

From one side, Stevens (1951) has defined measurement simply as the "*assignment of numerals to objects or events according to rules*". In this case, "*measurement scales should be regarded as arbitrary formal schemas and adopted in accordance with their usefulness for describing empirical data […] Such assignment of numbers to sensations counts as measurement because it is consistent and non-random, because it is based on the matching operations performed by experimental subjects, and because it captures regularities in the experimental results […] these conditions are together sufficient to justify the use of a ratio scale for measuring sensations, despite the fact that sensations cannot be separated into component parts, or laid end to end like measuring sticks*" (Tal 2020).[27]

---

[26] A useful suggestion is to try to avoid relative measurement: in particular, percentile ranks, standard deviations, and shares of variance (National Research Council 2011).

[27] This position may be referred to as "operationalism", roughly meaning that (social) concepts (especially Ballung concepts) have empirical meaning only if and when they are defined by measurement operations. This position has encountered a number of critiques such as that operational definitions are very specific and do not exhaust meaning, the ambiguities surrounding the notion of operation, the overly restrictive operational criterion of meaningfulness, and the fact that operational definitions are not required for all useful concepts (see Chang 2009). This position is opposed to "realism" where measurement is a mere approximation of a "true value" that is independent by procedural operations. This process is aimed at gaining insights about real properties rather than

Box 2.4 Metrics of fake news

As explained in Box 1.2, the term fake news is very loose (a Ballung concept again). In order to make it operational, we drew up a list of all the online news considered to be false after an assessment carried out by Italian fact checkers (i.e., Butac, Facta, NewsGuard, Open, Pagella Politica). The database contains 350 fake news (period 2016–2018), belonging to one of the following news categories: current affairs, politics, foreign affairs, showbiz, and science. We then associated each news item with the day of publication, imposing t_0 (i.e., the date on which the event it is occurred) equal to zero. The figure shows the time distribution of the number of occurrences of the news among news outlets: the distribution is very skewed to the right (at t_0) and then quickly decreases to zero within a few days.

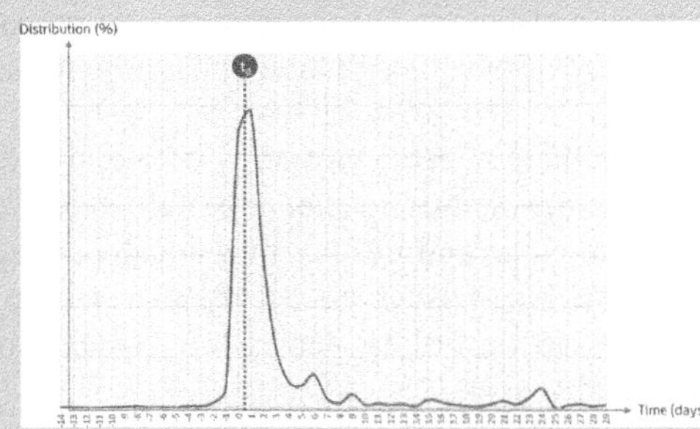

We then extracted a random sample of 350 articles from a database of all Italian online news sources (in the same period) to follow the same category distribution. We developed two metrics of news duration: the "average duration time", i.e., the average distance between the first and the last day on which a news item records at least one occurrence; the "average presence in the news", i.e., the average number of days, even if not consecutive, on which a news item records at least one occurrence. If we compare the two metrics of news duration, we notice that a fake news lasts, on average, a much shorter period of time: 6 days (compared to 30 days for real news), with an actual time of permanence of 3 days (20 days).

---

assigning values to objects. In this context, concepts such as measurement error and accuracy gain significance.

Metrics of news: real news vs. fake news

	Average duration time (days)	Average presence in the news (days)
News	30	20
Fake news	6	3

On the other side, Krantz et al. (1971) have moved away from the fundamental theory of measurement by proposing the Representational Theory of Measurement (RTM) which remains the most prominent mathematical theory to date.[28] RTM defines measurement as the construction of mappings from empirical relational structures into numerical relational structures. An empirical relational structure consists of a set of empirical objects along with certain qualitative relations among them (e.g., ordering, concatenation), while a numerical relational structure consists of a set of numbers (e.g., real numbers) and specific mathematical relations among them (e.g., "equal to or bigger than", addition). RTM is a formal and axiomatic generalization of the old concepts of the measurement theory which relaxes some conditions (e.g., concatenation), developing new broader concepts such as the "additive conjoint measurement", that can be used to quantify attributes where it is not possible to combine the levels of the attributes using concatenation.

However, even RTM has been met with a share of criticism, since it advances an abstract conception of measurement that is not connected to empirical work as closely as it should be (Heilmann 2015). Moreover, even if RTM rejects the idea that additive scales are adequate only when concatenation operations are available, this has not entirely solved the dispute with psychometrists (on this issue see, among others, Vessonen 2017).

A related issue regards the (flow of the causal) relationship between measures and phenomena. Usually, we identify structural relationships among latent, unobserved phenomena by statistically relating covariation between the latent constructs and the observed measures of the latent constructs. In this case, if variation in an indicator M is associated with variation in a latent phenomenon X, then exogenous interventions that change X can be detected in the measure M. In other words, we measure unobserved phenomena throughout a "reflective" measurement, with a causality flow from

[28] RTM goes into great detail in clarifying the assumptions underlying the construction of different types of measurement scales. Each type of scale is associated with a set of assumptions about the qualitative relations obtaining among objects represented on that type of scale. From these assumptions, or axioms, the authors of RTM derive the representational adequacy of each scale type, as well as the family of permissible transformations making that type of scale unique. In this way, RTM provides a conceptual link between the empirical basis of measurement and the typology of scales.

the latent phenomenon (X) to the measure (M). This is so for many psychological metrics.[29]

In economics and sociological contexts, it is often the other way round: "*This is sometimes called a "formative" as opposed to a "reflective" model of index construction. Examples include an index of socioeconomic status, consisting of education, income, and occupation, and the consumer price index, which is based on a market basket of goods and services. The construct is in fact determined by or defined by the indicators that go into it. Typically, the indicators are simply unit-weighted, but in some cases they are weighted on the basis of theory, differential utilities, or other preferences (e.g., relative importance based on a community survey)*" (National Research Council 2011).

The difference between the two types of measurement lies in the nature of the phenomena: i.e., in a reflective model, the latent construct exists independent of the measures, while in a formative model, the latent construct is dependent upon operationalist instrumentalist interpretation by the researcher[30]; as said, in the direction of causality (from the phenomenon to the measure for reflexive model and the other way round for constructive model); and in the characteristics of measures: "*In a reflective model, change in the latent variable must precede variation in the indicator(s). Thus, the indicators all share a common theme and are interchangeable. This indicator interchangeability enables researchers to measure the construct by sampling a few relevant indicators underlying the domain of the construct [...] Inclusion or exclusion of one or more indicators from the domain does not materially alter the content validity of the construct. However, the situation is different in the case of formative models. Since the indicators define the construct, the domain of the construct is sensitive to the number and types of indicators representing the construct. Hence, adding or removing an indicator can change the conceptual domain of the construct*" (Coltman et al. 2008).

Finally, the two approaches need different systems of (statistical) validation. For instance, within the reflexive framework, a numerical scale of measurement can be employed, under the conditions of internal and external consistency. The numerical scale of a phenomenon (e.g., depression) can be developed provided it is validated by an internal consistency analysis (e.g., using measures such Cronbach's "α") and a process of external verification with respect the latent construct (i.e., analyses that prove that the numerical scale is a good representation of the phenomenon investigated through, for instance, a Rach model; e.g., in the case of depression, clinical studies have shown that the MFQ robustly represents the pathology, and on the basis of these specific cut-offs of the index have also been defined, also broken down by age and gender, see Box 2.3).

---

[29] "*a test is valid for measuring an attribute if and only if a) the attribute exists, and b) variations in the attribute causally produce variations in the outcomes of the measurement procedure*" (D. Borsboom 2005).

[30] This difference mimics the distinction between realist and operationalist/conventionalist views (see above).

To conclude, in social sciences, measures are developed either in contexts in which a direct isomorphic or homomorphic mapping between phenomena and measurement scales can be defined, or thorough a model-based approach. In this latter case, the construction of measures can follow a reflexive or formative approach.

2.2.1 Measures of Threats to Journalists

After having defined and classified the phenomenon of threats to journalists (see par. 2.1.1), the different possible events can be measured along two dimensions (see Table 2.2).

First, events (i.e., threats) may or may not occur (i.e., "occurrence"). A measure, therefore, could be of a logical ordinal nature (Yes/No) and can be summarized through a binary variable (0/1), which assumes a value equal to 1 when the event occurs and is 0 in the opposite case. In addition, it is possible to quantify the

Table 2.2 Measures of threats to journalists (timeframe: 12 months)

Category	Measure #1	Measure #2
	Occurrence	Frequency
Total threats	At least one (0/1)	–
Traditional	At least one (0/1)	–
Aggression	Yes/No (0/1)	Number (0, 1, 2, 3, …)
Warning	Yes/No (0/1)	Number (0, 1, 2, 3, …)
Damage to personal property	Yes/No (0/1)	Number (0, 1, 2, 3, …)
Web	At least one (0/1)	–
Diffusion of false personal information	Yes/No (0/1)	Categorical (0-"never", …, 4-"every day")
Offence	Yes/No (0/1)	Categorical (0-"never", …, 4-"every day")
Sexual offence	Yes/No (0/1)	Categorical (0-"never", …, 4-"every day")
Threats of violence	Yes/No (0/1)	Categorical (0-"never", …, 4-"every day")
Legal actions	Yes/No (0/1)	–
Obstacle to information	Yes/No (0/1)	Number (0, 1, 2, 3, …)

"frequency" of the event. The intimidations have been measured along two scales according to their type: "traditional threats" and "legal actions" according to the number of occurrences (i.e., numerical scale), while web threats, potentially very frequent, along an ordinal (categorical Likert 5-point) scale: "never" (0), "sometimes" (1), "once a week" (2), "several times a week" (3), "every day" (4).

In the case of threats to journalists, occurrences and frequencies are measured over a 12-month period.

To sum up, after having classified the phenomenon of threats, we construct simple and clearcut measures to quantify it numerically (i.e., mapping). The numerical scale is ordinal and, for some of them, additive. The measurement approach is reflexive.

2.3 Measurement of Social Phenomena

Once the perimeter of the phenomenon and the metrics be used to measure it have been defined, it is necessary to proceed to the measurement step, i.e., the one in which procedures for gathering data for measurement purposes are established and data are collected. Note that this step is sometimes called "procedures" (or, more generally, "method") to differentiate it from the definition of measurement system and scales. However, I prefer to call it measurement because it is only at this point that the previous steps are accomplished, and the measurement of social phenomena takes place.

While the measurement methodology (and the related problems to be addressed) is very complex,[31] the main issues are rather simple.

When we aim to measure social phenomena, we must first identify a reference population to be analysed. This can be the whole population, a part of it (e.g., children or the elderly), some particular professional categories (e.g., journalists), social groups (e.g., party members), nations (all or only some of them, e.g., OECD member countries), enterprises (all or only some of them, e.g., selected by turnover or by industry through sectoral classifications such as NACE), institutions (e.g., antitrust authorities), etc.

Then we must determine the procedures to use for gathering the information we needed to calculate the metrics we have previously identified. Data can be collected from external sources (open or at least accessible to the researcher) and/or through direct measurement procedures, such as surveys and experiments. Data on citizens may come from sources such as census, surveys, experiments, and more recently data from the internet and related digital services (e.g., social media, e-commerce, web navigation), while business data come from sources such as balance sheets, professional databases, surveys.

[31] As for the definition of metrics, there is a huge literature especially of a statistical nature on methods for measuring social phenomena, to which we refer (you find some references in the bibliography at the end of the book).

Each of these external data sources (e.g., census, professional databases, social data) or direct data gathering procedures (e.g., surveys, experiments) is associated with a possible target of the reference population. The census includes (almost) the whole population, while a survey or an experiment may be carried out only on a part of it.

Therefore, the choice of measurement procedure affects which part of the reference population can be analysed: from the whole to a (small) subset of it.

Box 2.5 Measurement of the disinformation phenomenon

To measure the disinformation phenomenon, we consider two metrics: (i) the overall proportion of disinformation over the total online news production; (ii) the overall audience reached daily by sources of fake news (i.e., subjective approach, see Box 1.4). We proceeded as follows. We identified, from independent fact-checkers (i.e., Butac, Facta, NewsGuard, Open, Pagella Politica), a list of all detected disinformation sources that varies over time and includes both websites and social networking pages. It should be noted that sources of disinformation are generally fragmented and open and close very frequently. For this reason, secondly, we calculated the overall production of online news articles and their reach of all sources of online disinformation net of duplication. Finally, the % of disinformation production has been calculated over the total news online information (i.e., 37,654,333 news documents from April 2016 to May 2020), while audience reach has been estimated based on both site and in-app consumption.

The procedures to measure the production of news documents (both fake and real), are based on a database of 2,019 Italian online news outlets, which represents the universe of all national online news sources. The active reach of both news and false news outlets has been calculated on the basis of AudiWeb data, a web analytics organization that provides data on the use of Italian websites based on a combination of panel-based and server-side measurement.

So, we can assess that, in four years (2016–2020), the average proportion of disinformation over the total online news production has been equal to 5.9%. This value is rising and varies substantially according to the political cycle (Fig. a).

a): Production of disinformation (%)

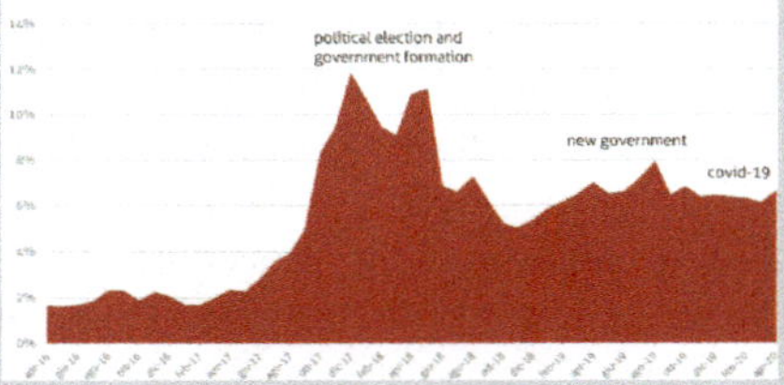

b): Consumption of disinformation (%)

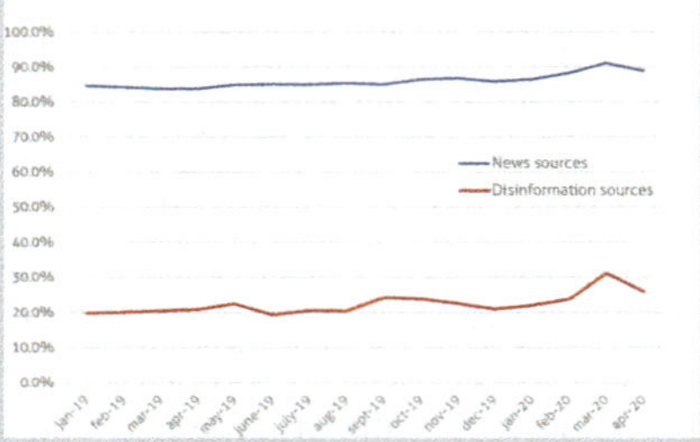

Note (a) % of online disinformation as a percentage of all online news production in Italy (2016–2019); (b) active reach of online news and disinformation sources in Italy (2019–2020)

As to news consumption (Fig. b), the overall audience of disinformation sites is lower than that of news outlets. However, this value is significant (ranging from 20 to 30% of all active users) and growing. During the first lockdown period, the audience of disinformation sources reached its peak: almost 14 million individuals (i.e., 31.1% of active users) have visited disinformation websites and social network pages.

The Internet has become a basic means for generating and a major source for collecting socio-economic data. In Table 2.3 new internet sources of data are listed.

Despite the profusion of massive data linked to new internet sources, it is still very rare to have information on the whole reference population, especially when this latter is large.[32] Instead, a subgroup of subjects is identified and analysed to represent the reference population. We call this group "sample". Methods for drawing a sample representing the whole reference population are dealt with by inferential statistics.

Two important concepts apply to a sample: randomness and representativeness. Because getting a random sample is very complex, randomness is difficult to achieve. Even if one manages to draw a random sample (e.g., in surveys and experiments),

[32] For example, in an analysis of the effect of regulation on addressing market failures in the agri-food sector, Castriota and Delmastro (2015) gathered detailed dynamic information (over more than 30 years) on all Italian wine denominations of origin from official databases. In this case they had data on the entire reference population.

Table 2.3 Sources of socio-economic Big Data

User's purpose	Description	Examples of sources
Information search	The user aims to find information about a topic of his interest. Data is actively generated	Search engines, Google Trends
Transactions	The user interacts with an individual and/or machine to achieve an agreement in which the user demands and obtains a product or service in exchange for a financial or non-financial compensation	
• *Financial transactions*	Event in which the user makes a payment to obtain a product or service	E-banking, e-commerce, urban sensors
• *Non-financial transactions*	Event in which the user provides the counterpart with required information to obtain a product/service	E-government, e-recruiting
Information diffusion	The user aims to spread information or knowledge. This includes marketing purposes, in order to establish a public image of the user or the agent he represents. Data is actively generated	Corporate websites, apps, Wiki pages
Social interaction	The user wants to share information, opinions and ideas with other users. Data is actively generated	Social Networking Sites, opinion platforms, blogs
Non-deliberate	The user does not pursue to generate data with action, but data are generated by the use of some means. Data is passively generated	
• *Usage*	The simple fact of using any device generates data related to how, when and where an action has been done	Web cookies, Internet Protocol, Sensors for self-tracking
• *Location*	The use of mobile phones generates data particularly related to the position of the user	GPS, GSM, Call Detail Records, Bluetooth, WiFi Points

(continued)

Table 2.3 (continued)

User's purpose	Description	Examples of sources
• *Personal*	Personal data (e.g. age, gender) is generated consciously (e.g. filling a form to complete a purchase) or unconsciously (e.g. data about the type of information we look for) as a consequence of using any device/tool to achieve a purpose	Forms, profiles, type of searches or purchases

Source D. Blazquez and J. Domenech (2018)

many subjects who have been selected may not respond and must be replaced, in a cascade process that inevitably moves the final sample away from the random one. This issue is more acute when information is personal (such as those about political preferences) or confidential (for companies, institutions), and sample subjects either avoid answering or provide misleading information.

Furthermore, to draw a random sample we need to do this from the reference population, so we must have basic information about the latter (e.g., from census data).[33] And this is not the case in many instances.

The representativeness of a sample is often associated with the sample's distribution with respect to a number of control variables (e.g., age, gender, socio-economic status and location where individuals live). However, even if the sample may mimic the universe with respect to the control variables, this is not sufficient to represent it with respect to the aspects that are under investigation. For example, even if there is a similar percentage of married individuals in the sample as in the reference population, it is by no means certain that, in the absence of sample randomness, the married individuals in the sample can represent those in the reference population. This phenomenon is called "sample selection bias".

Box 2.6 Measurement of depressive symptoms

As illustrated in Box 1.3, the use of psychometric methods is widespread in the analysis of psychological symptoms. In particular, the MFQ has been developed and used to detect depressive symptoms and anxiety disorders.

To this end, Delmastro and Zamariola (2020) explored the effects of COVID-19 on mental health. We administered the SMFQ (the short version of MFQ) to a random and representative sample of the Italian population. The study was conducted right after the first lockdown phase, in June 2020, to collect

[33] In social studies, the response rate is sometimes requested (with minimum thresholds, usually 20–25%, considered as necessary). However, this information is uninformative, if not completely meaningless, if one does not know how the initial sample was drawn.

the immediate reactions of the Italian population to the emergency. The final sample was composed of 6,692 Italian individuals, representative of the Italian population in terms of age, gender, and geographical area. More specifically, the sample design and the stratification were based on the following variables: (i) age; (ii) gender, (iii) geographical breakdown, and (iv) education. As a preliminary remark regarding internal consistency, it should be noted that Cronbach's alpha—i.e., a measure of the reliability of the questionnaire—is equal to 0.91 for the overall sample, which is well above the threshold of acceptance (i.e., 0.70; see Cortina 1993).

The first step was to measure the distribution of the SMFQ score among Italian population, just after the lockdown period (see Figure). As we would expect, the total distribution of SMFQ score is asymmetric and skewed to the right (also by skewness tests), with mean of 5.2. It presents a mode near 0, and a slight bump around 12. If we define a cut-off at 12 that may indicate the presence of depressive symptoms in the respondent (see again Box 1.3), then, in June 2020, 14.4% of the Italian population lied above this threshold.

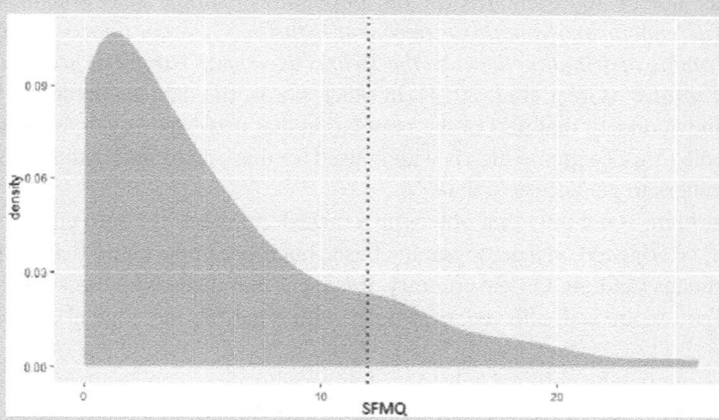

Distribution of SMFQ Scores on Italian population

Source Delmastro and Zamariola (2020)

Although sampling can in practice lead to multiple errors in the analysis of the reference universe, applied research does not have many alternatives, except as mentioned in the case where data on the universe were available (see Box 2.5). In this context, the advent of big data has completely changed the framework: "*Yet sampling comes with a cost that has long been acknowledged but shunted aside. It loses detail. In some cases, there is no other way but to sample. In many areas, however, a shift is taking place from collecting some data to gathering as much as possible, and if feasible, getting everything: N = all*" (Mayer-Schonberger and Cukier 2013).

However, the use of big data has not solved the problem.

Take the case of electoral forecasts which is very illustrative for our purposes. There is already a huge literature using social media data to predict citizens' electoral choices (see Bovet et al. 2018). This has stemmed both from the low-cost availability of social data, which provide insights into many aspects of people's political ideology, and from disaffection with traditional election polls that have evidenced major predictive errors in key elections (e.g., Brexit 2016 US Political Elections). However, the use of social data has greatly improved the knowledge about many aspects of human thinking, including ideological and political thinking (even providing massive tools to influence it[34]), rather than providing robust electoral predictions.

The reasons behind this outcome offers useful information on applied social science methodology.

First, n (i.e., the size of the sample) is always lower than N (i.e., the size of the reference population). Big data is never the whole. But the perception of having a huge number of data often overshadows the fact that the sample, although very large, may not be representative of the population (sample selection bias never dies!): "*The conclusion from this experiment is the need for continuous monitoring of demographic distribution between the Twitter users, and careful adjustment of the predicted results*" (Grčar et al. 2017). In other words, the data available are limited because they concern that part of the population that uses, and is active on, specific social media (e.g., Twitter, which is widely used for these analyses because it is easier for researchers to get access to data[35]).

The problem is not only that of having a biased sample of the reference universe, which can be adjusted on a demographic basis, but more importantly that the sample is by no means random. On the contrary, it is clear that those who use social media, such as Twitter, are self-selected, especially with regard to ideological and political issues.

Can we predict the political behavior of those who are not on Twitter (or any other social network) from those who actively use it? Probably not. Not only because there is a demographic bias between Twitter users and the voting population, but also because, all things being equal (e.g., age, gender, location), Twitter active users behave differently from the rest of the population (e.g., they are more engaged).

There is another issue that also gains relevance from an epistemological point of views. Social media data often provide insights into some important aspects of human thinking, but not directly into the behavior under analysis. It is rare that we gather voting intentions and behavior from browsing data on a social networking

---

[34] In a recent report eMarketer shows that worldwide 57% of major research buyers/suppliers make use of nonconscious online market research techniques such as eye tracking (38%), behavioral economic models (32%), facial analysis (24%), applied neuroscience (20%), biometric response (16%) (see eMarketer, *Biometric Marketing 2019: Revolutionary Personalization Tool or Targeting Gone Awry?*, October 2019).

[35] Note however that in October 2020, Twitter was the third largest social platform in the US with a 14% share (source: StatCounter).

site, except for that (small) part of users that is more ideologically committed. This problem is even more acute in contexts where the political offer is highly fragmented and/or where there is high voter volatility (as has been the case in Europe in recent years).

The underlying issue is that the data we have available on social networks (or more generally through big data) are often different from what we would like. The effect on social science is the risk of it being driven by the data available, both in the objectives of the research and in the methods by which we measure phenomena (i.e., classifications, metrics and measurement methodology).[36]

In short, while the approach to social science with traditional methodologies (e.g., surveys, experiments) has some significant methodological shortcomings in the phase of measuring phenomena but leaves the decision on the structure of the study to the researcher, the use of big data has often resulted in a subordination of the research to data. And we are back to the end of theory (see above).

This is not to say that data from non-traditional internet sources are not fundamental to the new social science,[37] but that they must nevertheless be traced back to a robust methodological structure in which the three steps we have described in this chapter are addressed.

We are therefore back to the starting point. Either we get (directly or indirectly) information on the entire reference population, or the problems of randomness and representativeness of the sample remain to be addressed and discussed. There are, of course, statistical, and procedural techniques to deal with these issues (e.g., a sample can be incentivised) but it is fair to recognise that none of them completely solves the problem.

A related aspect concerns the veracity of the information we have available. After all, a subject may respond untruthfully to the information we ask for (not only an individual, but also a company, an institution or even a government), just as many of the behaviours subjects have on the internet may not accurately represent what they want to express or do in real life.[38] Again, there are statistic and procedural techniques to deal with these issues (e.g., there are methods to calibrate questionnaires and to assess consistency of responses), but once again they do not fully solve the underlying problem.

In sum, when analysing social phenomena, the aim is to describe a reference population. Either we have information on the whole universe, or we examine a representative sample of the universe. In practice, however, drawing a representative random sample of the reference population is very complicated and this process must be often regarded more as a goal than something feasible. This is especially

---

[36] Companies offering digital services (such as search, social, e-commerce) are well aware of this fact, so much so that many of them have developed partnerships with companies that hold traditional data, such as credit cards and traditional marketing companies.

[37] For instance, text data (see Gentzkow et al. 2019) is key to analyse many social phenomena (see Chap. 4).

[38] Note that the veracity of data is usually inversely proportional to the amount of data acquired. This is a standard result in big data analysis, but it is often neglected. Traditional structured data exhibit a much higher degree of veracity.

so if there is no detailed information about the reference population from which to draw the sample, or if the latter is not large enough so that the process of replacing non-respondents becomes difficult or biased. The increasing availability of a massive amount of digital data may suggest that we can get information on the whole population, but this is often not (yet) the case. Sample randomisation procedures based on digital data do not solve the problem as the source data often concern a biased subset of the reference population. Furthermore, data integration with traditional information (through ex-post weighting procedures) can improve predictive performance but does not entirely solve the problem since the subset analysed may (and very often does) behave differently from the rest of the population (which lies outside the scope of the data). Finally, behaviours observed in all measurement procedures (surveys, experiments, digital data, etc.) can be distorted because they may not faithfully reflect the behaviour of subjects in real life. There are complex statistical and operational procedures for dealing with these issues, but it is fair to admit that none of them completely succeeds. Measurement, therefore, remains an area where practice and knowledge of the phenomenon remain essential and where, in addition to statistical and operational techniques, a case-by-case analysis is required.

The next sections illustrate the challenges of measuring a specific case (i.e., threats to journalists) and the practical procedures put in place to address them.

2.3.1 Measuring Threats to Journalists

As there are no reliable data sources on threats to journalists (see also Sect. 2.4), we proceeded to carry out an ad-hoc survey. The survey conducted in autumn 2018 involved a large sample of Italian journalists: 1,872 individuals, equal to 6.2% of the reference population (i.e., Italian working journalists, that we estimated at 30,194 from data of the National Institute for Italian Journalists' Social Security, INPGI).[39]

The sample is not random. When analysing a population as specific as that of journalists, it is impossible to find an open access census register from which to draw a random and stratified sample.[40] This is a typical case in social science research.

So, we employed aggregate socio-demographic statistics (in terms of age, gender, location and income) from INPGI to calculate ex-post sampling weights.[41]

[39] I gratefully acknowledge INPGI, and in particular Mimma Iorio, for providing aggregate information and data on the distribution of journalists in Italy (by gender, age, professional status, region, and income).

[40] In Italy, the only registry from which a representative sample of active journalists can be drawn is that of INPGI but for privacy policy reasons it is not accessible, if not for more aggregate kind of information (see previous note).

[41] The weighting procedure was carried out in collaboration with the Italian National Institute of Statistics—ISTAT (Federico Polidoro is gratefully acknowledged). This procedure resulted in an estimation method stemming from the resolution of a constrained minimum problem that falls within a general class of estimators known as constrained weighting estimators. The weight to be given to the sample unit is obtained by means of a complex procedure that: (i) corrects the distorting effect

The size, both in absolute numbers (i.e., almost 2,000 individuals) and as a proportion of the reference universe (i.e., 6.2%), as well as the sample weighting process makes the available data statistically robust, or at least as robust as possible.

A computer-assisted survey information collection (CASIC) method was used to gather information. Italian journalists were contacted by email and their participation was also encouraged by all relevant associations in Italy (i.e., FNSI, Ordine dei Giornalisti, USGF and USIGRAI which are gratefully acknowledged). This process led to the most widespread analysis of journalism ever conducted worldwide on a single country.[42]

The questionnaire covered all aspects of the journalistic profession. This allowed to address multifaceted traits of a multidimensional phenomenon (see Chap. 3), and to further mitigate self-selection with regard to the phenomenon under investigation (i.e., threats to journalists). The part of the questionnaire regarding threats has been formulated according to the classification illustrated in paragraph 2.1.1 (see Table 2.1). The method is self-reporting that is the only one that enables to analyse the shadow part of this phenomenon, which in turn represents a significant component of all acts of violence and not only (see next section).

Table 2.4 presents the distribution of the universe of all Italian journalists and sample journalists by gender, age, and Italian working macro-region.

Although the sample is not random, it represents, both in absolute value and in terms of the distribution of socio-economic characteristics, a fairly faithful representation of the reference universe. Moreover, the applied operational procedures (e.g., wide-ranging survey, mobilisation of all components of the profession—both employees and freelancers) mitigated the problem of self-selection.

2.3.1.1 Summary Statistics

In 2018 (see Table 2.5), slightly less than a quarter of all Italian journalists (23.6%) suffered some form of intimidation (at least one threat among all types considered, see Table 2.1). Traditional threats represent the most widespread form of intimidation (14.1%), immediately followed by online threats (13.8%). More limited, at least in quantitative terms, is the incidence of legal complaints (10.1%) and actions aimed at hindering information (7.8%).

Among the traditional threats, verbal warnings are the most widespread (10.9%) and physical aggression the uncommon ones (1.4%). Similarly, insults are prevalent on the web (13.4%), while those of a sexual nature (2.4%), as will be illustrated later on, affect only a specific group of journalists (i.e., younger women, see Chap. 3).

In addition to the size of the phenomenon, Table 1.5 also provides information about the frequency of threats to journalists. As can be inferred from both the

---

of the total non-response due to the inability to interview some journalists; (ii) takes into account the knowledge of known totals of important auxiliary variables (i.e., age, gender, income and Italian macro-area of work) available from INPGI.

[42] See Agcom, *Observatory on Journalism*, November 2020.

Table 2.4 Socio-demo characteristics: reference population vs. sample

Socio-demo characteristics	Italian journalist	
	Universe (%)	Sample (%)
Gender		
Female	42	40
Male (and other)	58	60
Age		
<30	6	5
31–40	23	23
41–50	30	27
51–60	28	32
61–70	10	11
<70	2	2
Italian macro-region		
North-west	34	35
North-east	18	17
Center	29	28
South	14	1
Islands	6	5

Note Universe values are elaborations from INPGI data and refer to all active Italian journalists in December 2018; sample values refer to weighted percentages

Table 2.5 Summary statistics of measures of threats to journalists

Category	Occurrence	Frequency	
	(%)	(Mean)	(Mode 1–4)
Total threats	**23.6**	–	–
Traditional	***14.1***	–	–
Aggression	1.4	0.02	1
Warning	10.9	0.28	1
Damage to personal property	1.9	0.04	1
Web	***13.8***	–	–
Diffusion of false personal information	6.8	–	1
Offence	13.4	–	1
Sexual offence	2.4	–	1
Threats of violence	4.7	–	1
Legal actions	***10.1***	–	–
Obstacle to information	***7.8***	0.23	1

Note The table shows the mean values for the numerical variables, while the mode refers to values 1–4, since otherwise, due to the characteristics of the phenomenon, this index would be always equal to 0

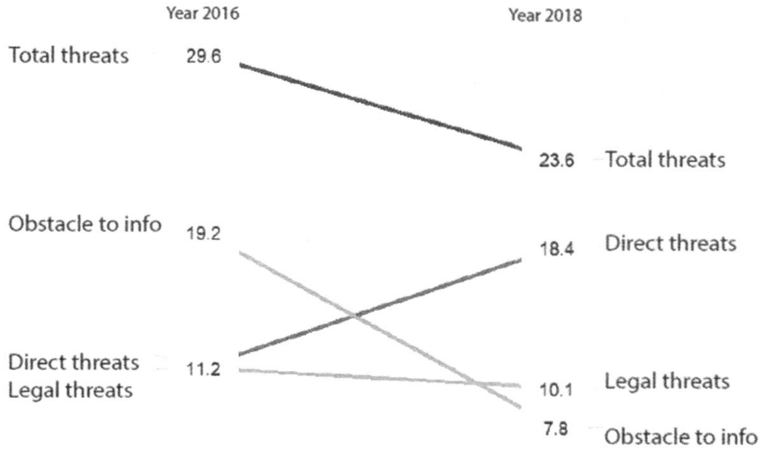

Fig. 2.1 Trend of threats to journalists in Italy. *Source* 2016 and 2018 ad-hoc surveys

average of the numerical measures (relative to traditional threats) and the mode (for values greater than category "never", i.e., "0"), the phenomenon of intimidation is widespread but occurs sporadically.

In other words, events in which a journalist receives more than one threat of the same type are quite rare. Consequently, the frequency does not provide much additional informative value (see also Chap. 3).

From a dynamic point of view, the use of a slope graph (see Fig. 2.1) allows to evaluate the trend of the phenomenon in a two-year period (2016–2018; for a description of the 2016 survey see Delmastro and Splendore 2020). During this period, there has been a general contraction in the phenomenon of intimidation perpetrated against Italian journalists, which, while remaining at very high levels, shows a clear downward trend (from 29.6% to 23.6%).[43]

This trend is due to the lower incidence of legal complaints (from 11.2% to 10.1%), and, above all, to the sharp drop in the number of cases of obstacles to information, whose incidence has been reduced by more than 50% (from 19.2% to 7.8%).

This last evolution seems to be largely due to the greater attention, both normative and social, to issues related to the transparency of administrative acts and documents (so-called national transparency strategy).[44] This refers in particular to the increased

[43] Note that this trend is confirmed by other external sources, which however cover only a limited, and less consistent part of the phenomenon (see Sect. 2.4.1). For instance, in the same period, the NGO Ossigeno per l'Informazione estimates that the number of intimidating acts has decreased by 34%.

[44] This consideration is also supported by international analyses on transparency that illustrate a condition of Italy that, starting from very backward potions, has started, since the last decade, a path of recovery: according to the NGO Transparency International, Italy, from 2012 to 2018, Italy has moved up from 72nd to 53rd position. Precisely in the period under consideration (i.e., 2016–2018), Italy has risen by 9 positions, and at the beginning of 2019 Italy was in 51st place (see https://www.transparency.it/).

availability of information and data, also in open data format, which is part of the new discipline of access to documents and public documentation. It was in fact in 2016 that in Italy the Legislative Decree no. 97/16 which protects the freedom of information and the right of access to administrative acts (so-called FOIA, Freedom of Information Act) was approved. From the results of the analysis, it seems that the introduction of a rule that recognizes the freedom to access information held by public administrations as a fundamental right, and, more generally, the adoption of a strategy of greater transparency, have contributed removing some of the obstacles to the exercise of the role as journalists of watching and scrutinizing political and administrative power. This evidence is in line with what has been found at international level in relation to similar regulatory interventions, and shows how much the legislative context can produce, or vice versa (as in this case) solve, a chilling effect, i.e., a deterrent effect on the exercise of fundamental rights, such as freedom of speech (see Chap. 3).[45]

2.4 Shadow Parts of Social Phenomena

The three-step procedure directs step by step the way in which social phenomena should be addressed. We have seen, for example, that the international classification of unemployment narrows the focus by leaving out discouraged individuals who are no longer looking for work. In addition, metrics may only cover a part of the analysed phenomenon, e.g., due to the availability of data only with reference to some traits of it. Finally, the measurement often concerns a part of the reference population. For all these reasons, the presence of a shadow (or hidden) part of the analysed social phenomena is very likely and widespread in social science.

[45] In this vein, Hazlett and Sosa (2007) verify the existence of a deterrent effect in the US context of the regulation of the radio market. The authors conclude that «*This is evidence that the old rules indeed provided a disincentive to broadcasting informational programs. The Supreme Court, if it is still looking for a "chilling effect", might carefully examine this experience in the radio broadcasting market*».

Box 2.7 COVID-19: measuring the shadow part

The measurement on the spread of coronavirus depends crucially on the detection system, which, in turn, is the result of: (i) the efficiency of the bureaucratic apparatus in detecting cases; (ii) the health strategies in carrying out (or not) tests (e.g., whether to do them also on the asymptomatic population or at least to those with mild symptoms); and (iii) political strategies. These factors affect the quality of data on confirmed cases, i.e., how much of the people who are actually positive to coronavirus are detected by official statistics. In other words, the hidden share of positives out of the total positive population (i.e., the reference population) depends crucially on the measurement system implemented by each country.

During the initial phase of the COVID-19 syndemic,[46] at the end of March 2020, Angelo Borrelli, the Head of the Italian Civil Protection Department, which in Italy is the institution in charge of managing the pandemic measurement, estimated that «*The ratio of a confirmed patient every ten [in Italy] is plausible*» (*La Repubblica*, interview of 24 March 2020).

At the end of 2020, during the second phase of the COVID-19 epidemic, some regions conducted mass testing in order to verify the actual percentage of positive cases (trying to reduce the hidden part as much as possible).

On October 31–November 1, Slovakia tested 3·6 million of its 5·5 million inhabitants using lateral flow tests (see Talha Burki, Mass testing for COVID-19, *Lancet News*, December 2020). Stefan Flasche, from LSHTM's Centre for the Mathematical Modelling of Infectious Diseases (CMMID), said: «*Slovakia's strategy identified more than 50,000 infections, many of which would otherwise have been missed*» (*London School of Hygiene & Tropical Medicine*, December 6, 2020). These results lead us to estimate that the hidden part of infections was about 48%.

South Tyrol, the Italian alpine province bordering Austria, followed in November 2020 (on November 20–22), by testing nearly two-thirds of its 5.5 million people in one weekend. South Tyrol identified 3,795 new infections, which implies that at that time the shadow part of infected was about 16%.

This is the case of acts of violence (i.e., the use of physical force or power, threatened or actual, that either results in or has a high likelihood of resulting in injury, death, psychological harm, or property damage), where many of the victims do not report the violent actions they have suffered. Since this component can be very significant in relation to the phenomenon, any analysis that does not take it into account is bound to fail. Moreover, the attitude to report this hidden component may

[46] Roughly speaking, a syndemic is the clustering of two or more concurrent or sequential epidemics in a population with biological and social interactions, which exacerbate the prognosis and burden of diseases. See Chap. 4 for a discussion.

vary over time (because, for instance, more victims may be encouraged to report the violence they have suffered).

Box 2.8 Measuring the size of the shadow economy

The shadow economy (also hidden, underground or black economy) is broadly defined as the economic activities that are hidden from public authorities and official statistics, usually to avoid taxation.[47] Assessing (dynamically) the size of the shadow economy is key for many reasons, e.g., because a prospering shadow economy impacts on crucial official indicators (e.g., unemployment, labor force, GDP, consumption), making them unreliable and undermining the quality of economic and social policies based on them.

However, "*a precise definition seems quite difficult, if not impossible, as the shadow economy all the time according to the principle of running water: it adjusts to change in taxes, to sanctions from the tax authorities and to general moral attitudes*".[48] Therefore, a first issue concerns the classification of the phenomenon itself (i.e., first step).

Moreover, the exact focus of analysis varies depending on the chosen metrics and method of measurement (i.e., steps 2 and 3), so that different approaches lead to different results.

In general, there are three methods of assessment of the size of the shadow economy[49]: (i) "the survey approach": direct procedures at a micro level that assess the size of the hidden economy at a particular point in time; (ii) "the modified currency demand approach": indirect procedures that employs macroeconomic metrics to estimate the evolution of the hidden economy over time; (iii) "the model approach": statistical tools to estimate the shadow economy as an unobserved variable.

Estimates of the shadow economy (period: 1999–2007) exceed, on average 50% of national income for countries with less developed economies (e.g., Azerbaijan, Georgia, Bolivia, Haiti, Myanmar, Nigeria, Panama, Peru, Tanzania, Thailand, Zimbabwe) but reach significant percentages also for most developed countries (e.g., estimated at around 25% for Greece, Italy, Spain and Russia, and above 10% for France, Germany, Japan and the United Kingdom).

---

[47] Schneider, F., & Enste, D. (2013). *The Shadow Economy: An International Survey*, Cambridge University Press.

[48] Schneider, F., & Enste, D. (2000). Size, Causes and Consequences, *Journal of Economic Literature*, 38, 77–114.

[49] Schneider, F., & Williams, C. (2013). *The Shadow Economy*, The Institute of Economic Affairs. See also, Kirchgässner, G. (2016). On Estimating the Size of the Shadow Economy, *CESifo Working Paper*, 5753.

So, even dynamic analyses carried out on official data may produce inconsistent results. Frequently official statistics show an increase in the number of cases of violence, whereas this trend is due to the emergence of the shadow component and not to a real increase in the phenomenon.

This issue is much more widespread than one might think, and, as explained in Boxes 2.7 and 2.8, concerns biological, economic and social phenomena such as the spread of the Covid-19 epidemic, and the assessment of the whole economic system.

2.4.1 The Shadow Part of Threats to Journalists

Acts of violence present a significant shadow component that goes well beyond official statistics (see Fig. 2.2). In this case, a survey-based methodology may be the one that ensures better coverage of the component not monitored by official statistics. Other methods are also feasible (see Box 2.8 in the case of shadow economy measurement methodologies), but in this case, the survey approach seems by far the most appropriate.

If we consider the results of the analysis conducted on Italian journalists (see paragraph 2.3.1) and we compare them with official statistics and administrative data, it turns out that, in the case of traditional threats (see Table 2.1 for definition and Table 2.2 for metrics), as many as 60% remain hidden because they are not reported to the police or other institutions (e.g., the Journalists' Association and other trade associations). Of the remaining 40%, only 2% are measured by official

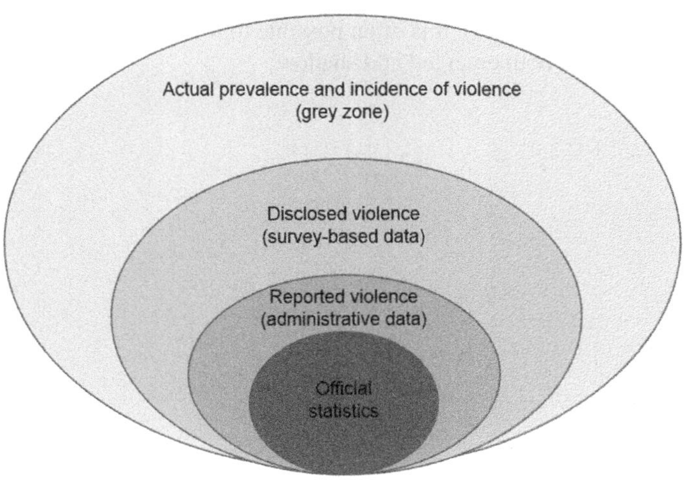

Fig. 2.2 Detection of acts of private violence: methodological issues. *Source* European Institute for Gender Equality

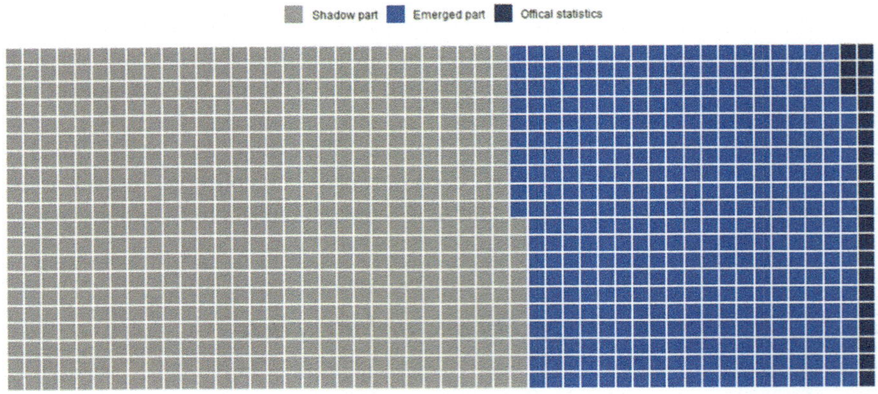

Fig. 2.3 Distribution of journalists who have suffered traditional threats by various categories. *Note* "Shadow part": intimidations suffered but not reported; "Emerged part": intimidations reported to various institutions; "Official statistics": data on intimidations to journalists by the Italian Ministry of Interior. *Source* Ad-hoc Survey, "Ossigeno per l'Informazione", "Centro di coordinamento per le attività di monitoraggio, analisi e scambio di informazione sugli atti intimidatori nei confronti dei giornalisti" of the Italian Ministry of Interior

statistics, while 38% are disclosed but, for various reasons, no trace remains at the official level.[50]

These values, which must be regarded as estimates, provide a measure of the size of the hidden component (which ends up being the greater part). After all, they highlight the crucial importance of assessing this component in the analysis of many social phenomena. By appropriately employing the three-step procedure (from classification to measurement), it is often possible to investigate all the components of social phenomena, both emerged and shadow.

[50] The analysis of the factors of a psychological (e.g., in terms of the "intrinsic motivations" of individuals in carrying out their tasks), sociological (regarding the role of the social context on individual actions) and behavioural (regarding the reactions of individuals following the manifestation of threats and intimidations of various kinds) nature is addressed in the following Chapters.

Chapter 3
Covariant Dimensions

In the previous chapter, we argued that social phenomena are vague and multifaceted to the extent that they require a methodology that rigorously addresses three fundamental steps: classification, metrics, and measurement. In this chapter, we explore another aspect of social phenomena: multidimensionality.

Social phenomena co-evolve and therefore cannot be studied in isolation.

Poverty, for instance, is a vague and multifaceted concept (i.e., "Ballung concept", see Sect. 1.1) in which operationalization inevitably depends on its classification and measurement (in terms of both metrics and detection procedures). At the same time, poverty is closely linked to other social phenomena ranging from unemployment to the social context, up to psychological traits. The concept of poverty itself can differ depending on these factors; for example, the socio-economic context influences not only the identification of poverty but also the perception of it.

Moreover, multidimensionality in social sciences is often associated with multidisciplinarity (in Chap. 4, we will discuss knowledge integration). Each social phenomenon co-variates with other aspects of various kinds: social, but also economic, demographic, psychological, up to physical and meteorological elements.

A large and growing part of the analysis of economic (e.g., purchase of goods and services), political (e.g., electoral decisions) and social (e.g., information choices) decisions is based on concepts, metrics, and experiments from cognitive psychology and neuroscience. The very foundation of choice theory, with its reliance on the rationality hypothesis, has been challenged by these scientific fields. As a result, nowadays it is questionable to analyze these phenomena without the use of tools and concepts deriving from psychological and neurological sciences.

Multidimensionality and complexity are, thus, two fundamental traits of social analysis and must be tackled through appropriate methodologies.

The step after the measurement of a phenomenon (see Chap. 1) is therefore the selection of the factors that covariate with the phenomenon itself (see Sect. 3.1). Then, these factors must be measured. In addition to the issues addressed in the previous chapter, here we illustrate the benefits of using indicators, both simple and composite,

M. Delmastro, *On the Measurement of Social Phenomena*,
SpringerBriefs in Political Science,
https://doi.org/10.1007/978-3-030-77536-0_3

that go well beyond rough data (see Sect. 3.2). Unsurprisingly, the increasing availability of data has been paralleled by the development of new techniques for defining statistical indicators as well as a massive literature on the subject. Finally, it is necessary to quantify the relationship between the identified variables (see Sect. 3.3). The advent of big data analyses has also resulted in a shift in emphasis from causal relationships to correlations. This is a philosophical debate going back centuries over causality and relations. In this chapter, we briefly summarize the merits and demerits of the two approaches, which are not only based on a different underlying philosophical perspective (often implicit in the analysis), but also have implications in the design of the research, in the statistical analysis and, above all, in the ultimate goal of social science.

3.1 Selection

The selection of covariant dimensions is key in shaping the research design. If the three-step procedure sets the domain of analysis of a social phenomenon, the selection of covariant dimensions sets the stage in which the phenomenon is investigated. On the one hand, selection should cover all relevant dimensions; on the other hand, this process may be limited by several factors including the availability of data on some dimensions.

Box 3.1 Covariant dimensions of profitability: the organizational structure

In industrial organization, firm profitability has been associated with many dimensions such as demand structure, market competition, business strategies (including advertising), and innovation (e.g., R&D and patents). Another widely debated dimension concerns the organizational structure of companies. However, this concept is rather vague and difficult to operationalize, except through rather discretionary classifications (e.g., U-form, M-form).

A new line of research has therefore attempted to analyze organizational structure by identifying some fundamental traits that can be measured objectively, and then verifying their impact on the profitability of companies (see Colombo and Delmastro 2008 for a survey). To this end, the organizational structure has been analyzed through an organizational chart based on which a series of metrics could be measured. These are: (i) the number of hierarchical levels that compose the organizational structure (i.e., the depth of the hierarchical pyramid); (ii) the average number of subordinates for each hierarchical level (i.e., the span of control); and (iii) the allocation of decision-making power (i.e., at what level formal and real authority concerning a range of strategic and operational decisions is assigned).

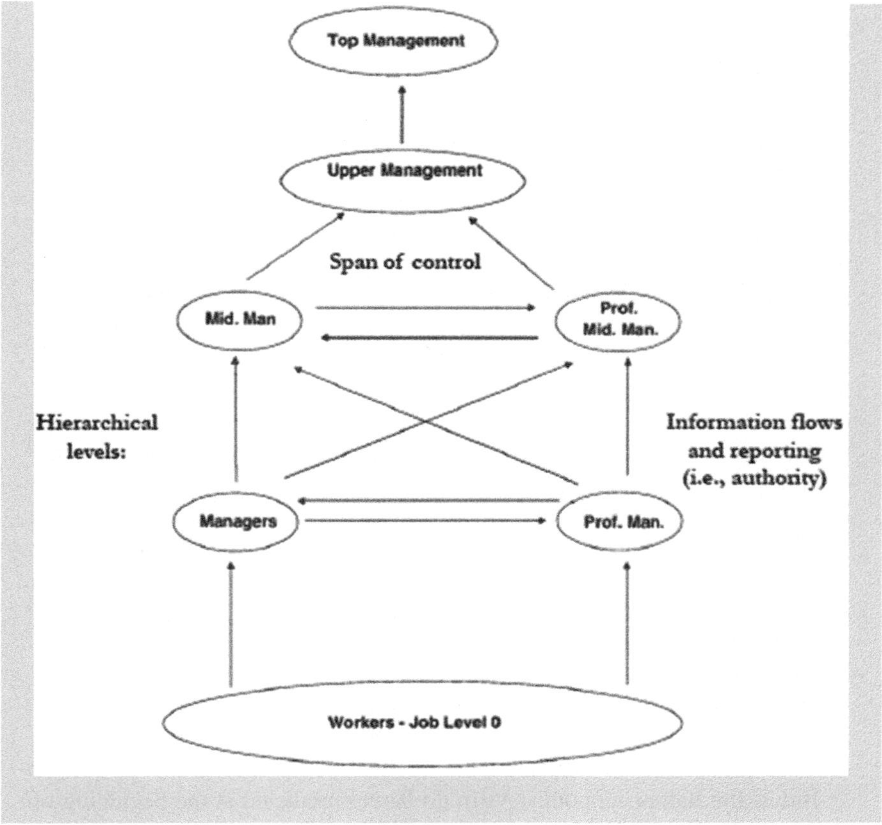

This process may take place at a theoretical level (i.e., starting from a reference model, and selecting those variables that summarize it), and/or at an empirical level (i.e., starting from a wide and heterogeneous dataset of available variables and then testing, possibly in an iterative way, the significance of their correlations with the phenomenon under study).

There exists an extensive literature (mainly statistical) on methods of variable selection. For instance, in the case of big data analysis, the selection of the variables (and also of the model, see Sect. 3.3) is often automatic and iterative (see Liu and Motoda 2012). These methods allow to find correlations between variables, even unexpected ones compared to theoretical models. However, these methods often raise issues, among others, of spurious correlations.[1]

[1] "*When the number of data points increases by orders of magnitude, we also see more spurious correlations - phenomena that appear to be connected even though they aren't*", Mayer-Schönberger and Cukier (2013). For a variety of examples of spurious correlations, see Vigen, T. (2015). *Spurious Correlations*, Hachette Books.

Box 3.2 Covariant dimensions of individual choice: rationality vs. emotions

For centuries, choice theory has been based on Descartes' principle of individual rationality (so-called classical theory of choice). Rationality is based on the Aristotelian idea of a calculating capacity and is understood as an individual's ability to optimize a utility function derived from axioms imposed on the consumer's own preferences (such as completeness, reflexivity, and transitivity). Classical choice theory developed formally, gradually approaching the domains of mathematics and statistics and building an axiomatic approach (see the theory of expected utility of Von Neumann & Morgenstern). In this context, choice decisions were mainly dependent on the information set of individuals, so that information asymmetries could lead to market failures. Public policies were therefore oriented towards forms of transparency and resolution of market failures through fair signaling instruments (e.g., in agri-food markets, labelling and designations of origin).

However, many of the violations of the axioms on which the theory of choice was based found on an empirical level needed a solution (e.g., the well-known Allais paradox).

A first development concerned the observation of the limitation of human cognitive processes, with the introduction of the concept of bounded rationality (Simon 1972[2]). In this sense, choice theory had to consider limits related to the human capacity to process information, essentially questioning the principle of maximization and introducing new cognitive dimensions.

But as the Italian economist Vilfredo Pareto predicted at the beginning of the twentieth century,[3] the theory of choice had to take into account previously neglected psychological dimensions. Based on considerations of cognitive psychology and neuroscience that stem from Hume,[4] the concept of rationality has gradually been abandoned, leading to new scientific fields (e.g., behavioral economics, experimental economics, neuroeconomics), and other dimensions

---

[2] "*Rationality can be bounded by assuming complexity in the cost function or other environmental constraints so great as to prevent the actor from calculating the best course of action. Limits on rationality stemming from this source have not been prominent in classical theories of rational behavior*". Simon (1972). Theories of Bounded Rationality, in McGuire, C. & Radner, R. (eds.), *Decision and Organization*, North-Holland Publishing.

[3] "*There may come a day when we can infer from the principles of psychology the laws of social science*", Pareto, V. (1906). *Manuale Di Economia Politica*, Società Editrice.

[4] "*Hume did just that, redirecting human thinking about the nature of mind to centuries to come*". Gazzaniga, M. (2019). *The Consciousness Instinct: Unravelling the Mystery of How the Brain Makes the Mind*, Farrar, Straus, and Giroux. For a discussion of the approaches stemming from Descartes (i.e., classical theory of choice) and Hume (i.e., cognitive theory of choice) see Damato, A. (1995). *Descartes' Error: Emotion, Reason, and the Human Brain*, Putnam Publishing.

linked to the emotions of individuals have been introduced.[5] Accordingly, public policies are introducing new instruments of addressing market failures, which are meant to nudge individuals towards socially efficient equilibria.[6]

What needs to be stressed here is the relevance of this methodological decision and the need for its definition in an appropriate and knowledgeable way.

In this regard, it is important to raise some points. First, too much emphasis, as is often the case, on significant relationships between variables can deprive research of important contributions. In many cases, the lack of significant relationships between variables can be of great scientific value. For instance, in studies conducted on Italy during the COVID-19 epidemic emergency, it was found that, contrary to what might be expected, there was no relationship between anxiety (and news seeking behavior) and the spread of the epidemic across the territory (see Castriota et al. 2020; Delmastro and Zamariola 2020). These results suggested that even in places where the epidemic had not (yet) arrived, states of anxiety were widespread, and policies needed to take this into account. In essence, the absence of a strong territorial correlation was an important finding for research and policies. This shows that the selection of variables should be made based on a conceptual reference model.

However, the conceptual model is often an excessively narrow boundary. This can happen for various reasons. For example, in social sciences such as economics, a theoretical-mathematical reference model to be tested may often be required. However, constructing a mathematical model that incorporates many variables and achieves testable equilibria is very complex. Hence, there is a tendency both to make restrictive assumptions about the functions representing the phenomena being considered and to narrow the scope of the latter. Nevertheless, in social sciences, the perimeter is very wide and limiting the scope of research too much can lead to disappointing analyses and inconsistent results (see Box 3.2).

This issue of the perimeter of the analysis and the selection of covariant dimensions applies to many research contexts. Often the problem is that some of the factors that should be taken into account relate to different scientific fields and social sciences are still too confined within their enclosure. For example, medical anthropologists

---

[5] "*emotions such as fear, affection, and hatred explain most of the occasions on which people depart from rationality*", and "*To a psychologist, it is self-evident that people are neither fully rational nor completely selfish, and that their tastes are anything but stable. Our two disciplines seemed to be studying different species, which the behavioral economist Richard Thaler later dubbed Econs and Humans*". Kahneman, D. (2011). *Thinking, Fast and Slow*, Penguin Books.

[6] See Thaler, R., & Sunstein, C. (2008). *Nudge*, Penguin Books.

have recently shown that when referring to the spread of coronavirus and the Covid-19 disease, it makes more sense to refer to a syndemic[7] rather than a pandemic.[8] Broadening the scope of the analysis to other health (such as anxiety and depression) and social problems (e.g., marginalized social groups) not only requires broader analyses but also different policies to counter the epidemic (see Chap. 4).

These brief remarks show how the selection of covariant factors is a fundamental research moment. Using a purely statistical (or even automatic) approach risks depriving research of important results that would derive from the factual observation when, at a particular time or for a particular population, no significant relationships may emerge between certain variables. On the contrary, following an approach based on a conceptual model may limit the perimeter of the research to the scope of social science or a particular strand of it.

The advent of big data has revealed how social sciences have often excessively limited their scope. Finding the right balance within a methodologically grounded approach is the challenge of the new applied social science.

In Chap. 4, we discuss how the integration of knowledge between different fields of social science has turned out to be a necessary evolution. In the remaining part of this chapter, we examine the other steps that need to be taken once the covariant dimensions have been selected, i.e., the measurement (through appropriate indicators) of covariant factors and the statistical estimation of these relationships. Before proceeding further, in the following paragraph, we illustrate the selection of covariant dimensions in the case of threats to journalists (see Chap. 2).

3.1.1 Selection of Covariant Dimensions of Journalistic Activity

Turning to the case study of threats to journalists, the research strategy involved, already in the design of the survey, the identification of dimensions that could theoretically covariate with the phenomenon under investigation. Therefore, the research design implied, not only the three-step procedure on the phenomenon analyzed (see Chap. 2), but also the selection of covariant dimensions and the related steps of classification, metrics, and measurement of them.

---

[7] A syndemic is a "*population-level clustering of social and health problems. The criteria of a syndemic are: (1) two (or more) diseases or health conditions cluster within a specific population; (2) contextual and social factors create the conditions in which two (or more) diseases or health conditions cluster; and (3) the clustering of diseases results in adverse disease interaction, either biological or social or behavioural, increasing the health burden of affected populations*" (Singer et al. 2017).

[8] See Horton (2020).

Based on the existing literature on journalism[9] and threats to the profession, the following dimensions of professional activity were identified, and related information was gathered[10]:

- Role in the news media system
- Dangerousness of the type of activity
- Proximity to the news
- Motivations
- Dangerousness of the working territory.

Since each dimension is, in turn, multi-dimensional, this implies that the various traits of these dimensions must be also identified to proceed to the subsequent steps.

3.1.1.1 Role in the News Media System

An initial dimension concerns the role that the journalist holds within the news media system.

Under this domain two traits are relevant.[11] First, there is a distinction between who is an integral part of an editorial staff (i.e., an employee) and who, instead, works autonomously (as a freelance). In Italy, the former enjoy contractual and income stability and, very often, insurance coverage that covers professional risks. Conversely, freelances work under precarious conditions, and usually on a lower income.

Second, among those who structurally work within a news outlet, one can distinguish between (hierarchic) roles with different duties and responsibilities. At the top, there is the editor-in-chief who, by the Italian Penal Code (Art. 57), is liable for "omitted control". The editor-in-chief is also liable under the Italian civil law with the author of an offending article and both jointly and severally with the owner of the publication (according to Article 11 of the Italian Press Act). In this sense, the editor-in-chief is more exposed, at least from a legal point of view, to external intimidation (e.g., through pretentious lawsuits, see Chap. 2, Table 2.1).

3.1.1.2 Dangerousness of the Type of Activity

The journalistic profession differs considerably depending on the activity carried out by the journalist. Working on food topics presents different professional risks

---

[9] See Hanitzsch, T., Hanusch, F., Ramaprasad, J., & de Beer, A. S. (2019). *Worlds of Journalism: Journalistic Cultures Around the Globe*, Columbia University Press, and Splendore (2017) for Italy.

[10] Socio-economic factors (i.e., gender, age, income, education) must be added to these dimensions.

[11] The classification of this dimension is based on existing contractual categories (under the Italian national contracts of FIEG-FNSI, Aeranti-Corallo, and USPI-FNSI) in terms of both distinction between employees and freelances, and identification of the professional figures within a news outlet.

compared to carrying out investigations on crime organizations. Nevertheless, measuring these differences through appropriate metrics is far from straightforward.

The first step is to divide the dimension into traits that characterize it. Based on the relevant literature and available data, the dangerousness has been made to depend on the following two elements:

- Main topic(s) covered by the journalist (e.g., food vs. current affairs)
- Type of tasks performed by the journalist (e.g., writing opinions vs. investigative journalism).

In both cases, since objective metrics are not available, the two traits have been classified by defining the perimeter of possible occurrences (considering only those that are relevant for the analysis).[12]

3.1.1.3 Proximity to the News

A third dimension concerns the proximity of journalistic activity to the news. Proximity is a central aspect of the profession (also in the case of digital journalism, see Ahva and Pantti 2014[13]), long debated in literature and by practitioners.

What is most relevant for the analysis, proximity is an element that can affect the likelihood of a journalist being threatened. It is of course one thing to write a war report by remote (desk analysis), it is another to investigate on the same topics on the ground.

The proximity dimension has been decomposed according to the following three traits:

- Proximity of the news outlet (e.g., national vs. regional)
- Proximity of the activity carried out (e.g., field work vs. desk analysis)
- Proximity of content (e.g., national/international vs. local).

Again, the three traits have been classified defining the perimeter of any possible occurrences (considering only those that matter for the analysis).[14]

---

[12] The topics taken into consideration are: (1) agriculture, (2) art, (3) current affairs, (4) culture, (5) economy/finance, (6) environment, (7) fashion, (8) food, (9) foreign affairs, (10) gossip, (11) health, (12) housing, (13) motors, (14) politics, (15) school/education, (16) science, (17) social, (18) showbiz, (19) sports, (20) technology, and (21) travel. The tasks taken into consideration are: (1) writing articles, (2) blog/web content, (3) editing/titles/page layout, (4) investigations/reportage, (5) opinions, (6) press review, (7) radio and/or TV conduction, (8) radio and/or TV news services, (9) videos, (10) web news aggregation.

[13] Ahva, L., & Pantti, M. (2014). Proximity as a Journalistic Keyword in the Digital Era, *Digital Journalism*, 2, 322–333.

[14] News outlets have been classified in: (1) national, (2) multi-regional, (3) regional and (4) local. The activities that are relevant for the measurement of proximity have been classified into the following occurrences: (1) commentary, (2) field news retrieval, (3) interviews, (4) video, and (5) other activities. Finally, the contents covered by journalists have been distinguished between national (and/or international) and local.

3.1.1.4 Motivations

Another dimension concerns professional motivations. It is well documented that the motivational trait plays a fundamental role in the individual professional performance (in this sense, there is a broad literature in economics, psychology and sociology; for a review of the psychological literature see, for example, Cerasoli et al. 2014[15]). This applies even more to activities such as journalism where economic aspects ("profit maximizing behaviour") are often overwhelmed by more general considerations ("utility maximizing behaviour").

In these contexts, the role played by "intrinsic motivations", i.e. «*the doing of an activity for its inherent satisfactions rather than for some separable consequence... When intrinsically motivated a person is moved to act for the fun or challenge entailed rather than because of external prods, pressures, or reward*» (Ryan and Deci 2000).[16] While the extrinsic motivations derive mostly from monetary rewards, the intrinsic ones originate within a person, and are directed by personal interests and satisfaction in performing certain activities.

By following this methodological perspective, we have proceeded to identify all the main motivational traits that characterize the journalistic profession. To this end, in line with international literature (see The World Journalism Study[17]), 14 motivational factors ("items") were identified.[18]

3.1.1.5 Dangerousness of the Working Territory

A final dimension concerns the territorial context in which the journalist works. All other things being equal, working in a war zone presents a significantly higher level of danger than working in a peaceful context.

In Italy, the dangerousness of a journalist's working environment derives essentially from two factors: the presence of common crime, and the presence of organised crime. Consequently, having acquired information about the workplace of the journalists, we can associate each of them with indicators about the dangerousness of the working territory (see Sect. 3.2.1).

---

[15] Cerasoli, C., et al. (2014). Intrinsic Motivation and Extrinsic Incentives Jointly Predict Performance: A 40-Year Meta-Analysis, *Psychological Bullettin*.

[16] Ryan, R., & Deci, E. (2010). Intrinsic and Extrinsic Motivations: Classic Definitions and New Directions, *Contemporary Educational Psychology*, 25, 54–67.

[17] See https://worldsofjournalism.org/.

[18] These are: (1) Being a distant observer; (2) Report the facts for what they are; (3) Providing news on current affairs; (4) Monitoring political leaders; (5) Monitor businesses; (6) Setting the political agenda; (7) Influencing public opinion; (8) Being a proponent of social change; (9) Providing entertainment and amusement; (10) Providing the type of news that attracts the broadest audience; (11) Providing advice and guidance for everyday life; (12) Providing information that people need to make political decisions; (13) Motivating people to participate in political activities; (14) Letting people express their positions.

3.2 Indicators

In Chap. 2, the definition of measures of social phenomena was addressed. In this chapter, we discuss a related issue, which is the construction of both simple and composite indicators. The use of indicators has increased in recent years with the growing availability of data from different sources. Nevertheless, their use, although fundamental for measuring social phenomena, must be framed within a methodological framework.

There are many reasons for employing indicators, one of which is the possibility of temporal and spatial benchmarking (see OECD 2008). I think that the most important motivation is the need to measure social phenomena in a way that is not merely a consequence of available raw data.

For example, in the case of threats to journalists one can introduce, as a covariant factor, simple (dichotomous) variables of the working region. In this way, we can identify whether a journalist works in a particular region (e.g., Sicily) and measure the territorial distribution of threats. Nevertheless, this territorial distribution may be the result of several factors that we can only conjecture having used the mere information about the place of work. Alternatively, we can associate each area (e.g., county, region, macro-area) where the journalist works with a territorial crime indicator and check the correlation between threats and crime. In this second case, we have proceeded one methodological step further: we have made explicit the covariant dimension we wish to introduce (i.e., the presence of crime) and used an indicator to test a conceptual reference model.

The use of indicators, therefore, helps us not only to measure social phenomena more robustly (so that we may benchmark them both spatially and temporally), but also to correctly define the overall framework within which the phenomenon under investigation is examined.

Indicators may be simple or composite. Roughly speaking, simple indicators refer to measures of only a trait of a phenomenon, so that measurement follows the rules illustrated in Chap. 2 (see Sect. 2.2). Composite indicators refer to measures of two or more traits of a phenomenon. In this case, additional methodological issues must be addressed.

We can briefly summarise the procedural steps that should be followed in the construction of composite indices, some of which follow what has been outlined both in Chap. 2 and in the previous paragraph:

1. **Conceptual framework** (i.e., selection and classification). A conceptual framework is the first step to provide the basis for the selection and combination of indicators of single traits into a meaningful composite indicator. Of course, the framework must also be consistent with the phenomenon within which the analysis is carried out (e.g., within the analysis of threats to journalists, territorial indicators may provide information on common and organized crime). In any case, a clear classification of the traits under consideration must be provided (see Sect. 2.1).

2. **Measurement**. The various traits must be measured according to the three-stage procedure outlined in Chap. 2 (see Sects. 2.2 and 2.3). In other cases, data can be selected and gathered from external sources, but care must be taken on how these have been measured (for instance see the issues related to the presence of a shadow component in Sect. 2.4).
3. **Statistical and data analysis**. The starting point of the empirical analysis is the creation of a dataset, which contains variables from one or more sources.[19] The dataset is then transformed into a form that is commonly referred to as "tidy" (Wickham 2014).[20] The most common operational procedures involved filtering, transformation, aggregation, and sorting. From a statistical point of view, this phase involved activities such as: consistency checking (between internal variables and with respect to external sources), managing missing data, and the reclassification of open-ended responses. An exploratory multivariate analysis should investigate the overall structure and consistency of data and indicators of single traits.
4. **Normalization**. Indicators of various traits of a dimension should be usually normalised to render them comparable. In this activity, attention needs to be paid to extreme values (i.e., outliers) as they may influence subsequent steps in the process of building a composite indicator. In this vein, skewed distributions should also be identified and accounted for.
5. **Aggregation (and weighting)**. Indicators of various traits of a dimension should be aggregated (and weighted) according to the underlying conceptual framework. Correlation and compensability issues among indicators need to be considered and either be corrected for or treated as features of the phenomenon that need to be retained in the analysis. There are many statistical techniques that address the steps of normalisation and aggregation. In what follows we illustrate some of them (see Sect. 3.2.1).
6. **Exploratory data analysis (EDA)**. EDA has developed impetuously thanks to new techniques of data management, processing, analysis, and visualization.[21] Today, researchers (especially in the social sciences) still underestimate the importance of visual tools, associating them only with new methods of exposing results (explanatory data analysis). New EDA techniques imply a new way of approaching applied research and allow, where used appropriately, a higher degree of understanding of the phenomena under observation, indicating keys of interpreting results as well as new avenues of research.

---

[19] A dataset is a structured collection of data generally associated with a unique body of work. A database is an organized collection of data stored as multiple datasets, that are generally stored and accessed electronically from a computer system that allows the data to be easily accessed, manipulated, and updated.

[20] A tidy dataset satisfies the following conditions: (1) each variable forms a column; (2) each observation forms a row; (3) each type of observation unit forms a table.

[21] "*When we reason about quantitative evidence, certain methods for displaying and analyzing data are better than others. Superior methods are more likely to produce truthful, credible, and precise findings. The difference between an excellent analysis and a faulty one can sometimes have a momentous consequences*", Tufte (1983).

The field of research on composite indicators has developed enormously in recent years. The interested reader can refer to the relevant literature.[22] In the remainder of this paragraph we will provide some practical examples illustrating these steps in the case of threats to journalists.

Box 3.3 Indicators of the organizational structure: allocation of decision-making

As remarked in Box 3.1, recent literature on industrial organization has examined the organizational structure of firms and institutions through new measures of factors widely discussed in theoretical research. As some of these concepts are rather vague, their operationalization presents methodological challenges.

In particular, while the concepts of size of the organizational pyramid (which can be measured, for example, through the number of hierarchical levels) and span of control (the number of subordinates for each organizational position) can be measured objectively, that of allocation of decision-making power is more elusive.

Some measures, which concern top-level decision-making, can be constructed based on elements such as the delegated powers granted to apical positions such as the chief executive officer. However, these do not allow for the analysis of the dense network of relationships of a complex system, which as noted above can be portrayed as information and decision flows within a comprehensive organizational chart.

Alternatively, through a methodology of field analysis (through face-to-face interviews or by surveys) the decision-making hub of a set of strategic and operational decisions can be mapped. This requires the identification of the decisions which are relevant for the analysis. This choice depends on the reference unit of the analysis: e.g., plants, companies, conglomerates. Furthermore, the allocation of decision-making power may provide for either strict delegation to an organizational level (i.e., the delegation of real and formal decision-making), or the power of the superior to overrule the decision of the hierarchically subordinate level at any time (i.e., the delegation of real but not formal decision-making).

Once these steps are defined (i.e., classification of the concept of decision-making power), and information is acquired (through a specific methodology), the measure obtained is multifaceted: i.e., for each decision, whether operational or strategic, the hierarchical level to which the relative decision-making power is attributed is identified, as well as how decisions are taken (i.e., formal and/or real power).

---

[22] See Di Nardo et al. (2005), OECD (2008) and the Social Indicators Research Series by Springer.

The last step is to create a composite indicator measuring the overall allocation of decision-making power, and thus the centralization or decentralization of power within a reference unit (e.g., a company). This can be done through various statistical methods. One possible method is that of principal component analysis, and the use of the first component as a measure of allocation of decision-making power. This choice depends on the structure of the data as well as on the extent to which the first component achieves in terms of explained variance. For an analysis of these issues, see Colombo and Delmastro (2008).

3.2.1 Indicators of Journalistic Activity

In Sect. 3.1.1 we illustrate the selection of covariant dimensions of the threats to journalists and thus the underlying conceptual model. This phenomenon has been related to five possible dimensions of journalists' activity: (1) role in the news media system; (2) dangerousness of the type of activity; (3) proximity to the news; (4) individual motivations; (5) dangerousness of the working territory.

In addition, we have also defined and classified the various traits of these dimensions. The next step is to develop simple indices of individual traits and possibly composite indices of each dimension. Since the various dimensions present different characteristics, separate design procedures consistent with the analysed dimension are made.

The aim is to move from raw data of some important dimensions (e.g., in the case of the territorial context, dummies of the macro-area in which a journalist works) to indicators of the same. This is to obtain measurements that are both synthetic, robust, and possibly additive (see Sect. 2.2).

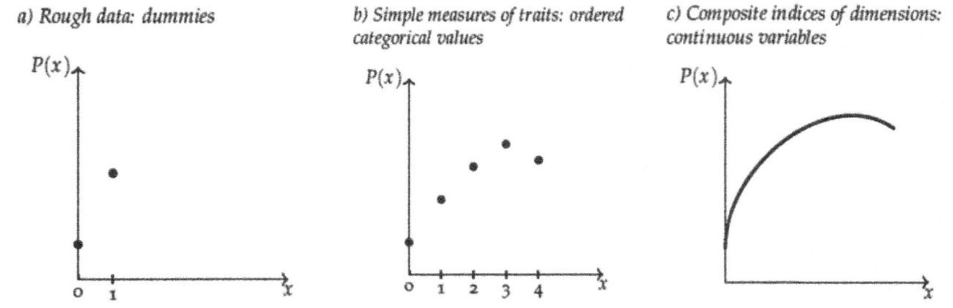

3.2.1.1 Indicators of Single Traits

As far as the **role of the journalist** within the news outlet is concerned, the information available provides synthetically an account of two traits: the structural affiliation (or not) to a news organization (employee vs. freelance) and, eventually, the role assumed by the journalist within the newsroom (i.e., trainee, junior and senior journalist, sub-editor, and editor-in-chief).

By the very nature of this dimension, the metrics can be aimed at capturing the occurrences of the different cases with respect to the universe of journalists, or, in the case of employees, to those who work stably within a news outlet. In this sense, dichotomous variables have been used that present a unitary value in the case of the occurrence of an event, with the two traits (employee/freelance and role in the newsroom: none—i.e., freelance—trainee, junior and senior journalist, sub-editor, editor-in-chief) that inevitably intersect one to another and must therefore be studied jointly (i.e., through the same metric system).

As to the **type of activity**, two main traits of the degree of **dangerousness** have been identified: (i) the topics(s) the journalists deal with; (ii) the type of tasks they carry out.

In both cases, since no objective metrics were available, the following steps have been taken. First, the main occurrences of each trait were identified: more than twenty in the case of topics, more than ten in the case of tasks (see Sect. 3.1.1).

Second, based on the relevant literature on the subject, two categorical systems have been constructed that ranked selected topics and tasks in terms of potential degree of dangerousness. The third step involved the engagement of a panel of experts and practitioners to whom the two ranked ordering were submitted, in an iterative way, asking them to confirm or rearrange the various topics/tasks according to their assessment of their degree of dangerousness (following the so-called Delphi method).[23]

The final results of this methodology are represented by two simple indices that measure the two analysed traits by using a comparable categorically ordered scale (0–3): i.e., the Dangerousness index of topics (DIT) and the Dangerousness index of tasks (DITk).

In sum, the two simple indicators of dangerousness of journalists' activity (DIT and DITk) are consistent metrics that contain as much informative value as possible, as derived from existing international and national evidence ("a priori" knowledge), as well as from knowledge dispersed among professionals in the field (Delphi method).

Since **proximity** is a physical concept, the construction of indices of traits of this dimension (i.e., proximity of news outlet, activity, and content) is easier because it presents characteristics of objectivity and is based on the expected distance.

[23] The Delphi method is a structured communication technique, developed as a systematic, interactive forecasting method that relies on a panel of experts. The main purpose is to exploit differentiated and dispersed knowledge for the analysis of technical topics. In this case, the panel was composed of 10 individuals, both journalists and academic experts.

As a result, three simple indicators have been constructed by evaluating physical distance on a categorical ordered scale: i.e., the Proximity outlet index (POI, from local, i.e., maximum proximity, to national outlet, minimum proximity); the Proximity activity index (PAI, from field work, i.e., maximum proximity, to desk activity, minimum proximity); and the Proximity contents index (PCoI, from local content, i.e., maximum proximity, to national/international content, minimum proximity).

Professional **motivations** concern psychological aspects which are measured by means of psychometric methods (see also Chap. 2). In this case, we employ a self-reporting method based on the major professional motivations ascertained in the international literature.[24] As illustrated above (Sect. 3.1.1), 14 motivational items were proposed with possible responses from journalists on a 5-point Likert scale, ranging from 1 ("Not important") to 5 ("Extremely important)".

To assess the internal coherence of the answers provided by journalists (in terms of expression of a latent construct), Cronbach's "α" index was used. For the 14 assessments expressed by journalists this measure is 0.75, which is above the threshold of acceptance (i.e., 0.70; see Cortina 1993).

At this point a process of "scale purification" has been implemented[25] (i.e., selection from the 14 items analysed) to achieve measures that ensure consistency, relevance and individual variance, on the one hand, and parsimony, on the other hand. For this purpose, the results of a related research work, that has conducted a statistical analysis of the 14 motivational items through the application of the so-called CUB models, has been used.[26]

Figure 3.1 illustrates how two items ("monitoring political leaders" and "providing information that people need to take political decisions") show a rather high heterogeneity (about 70% of the maximum) and tend to divide journalists into distinct groups. Moreover, these two factors represent fundamental motivational elements in the journalistic profession: the monitoring function of journalism over (economic and political) power, and the production of information relevant to citizens' decision-making (see AmericanPress Institute, *The Elements of Journalism*).

[24] In this vein, see for instance the "Intrinsic Motivation Inventory", i.e., "*a multidimensional measurement device intended to assess participants' subjective experience related to a target activity*". The IMI scale was created by Deci, Plant and Ryan; see Deci, E. (1971). The Effects of Externally Mediated Rewards on Intrinsic Motivation, *Journal of Personality and Social Psychology*, 18, 105–115.

[25] "*In the case of reflective measurement with multiple items, these items are expected to correlate, and, in order to improve psychometric measurement properties, the researcher may need to eliminate a number of them*", Jarvis et al. (2003).

[26] Capecchi, S., Piccolo, D., & Simone, R. (2020). Analisi dei Fattori Considerati Importanti dai Giornalisti nello Svolgimento del loro Lavoro, Università di Napoli "Federico II", mimeo. See, also, Capecchi, S., & Simone, R. (2019). A Proposal for a Model-Based Composite Indicator: Experience on Perceived Discrimination in Europe, *Social Indicators Research*, 141, 95–110.

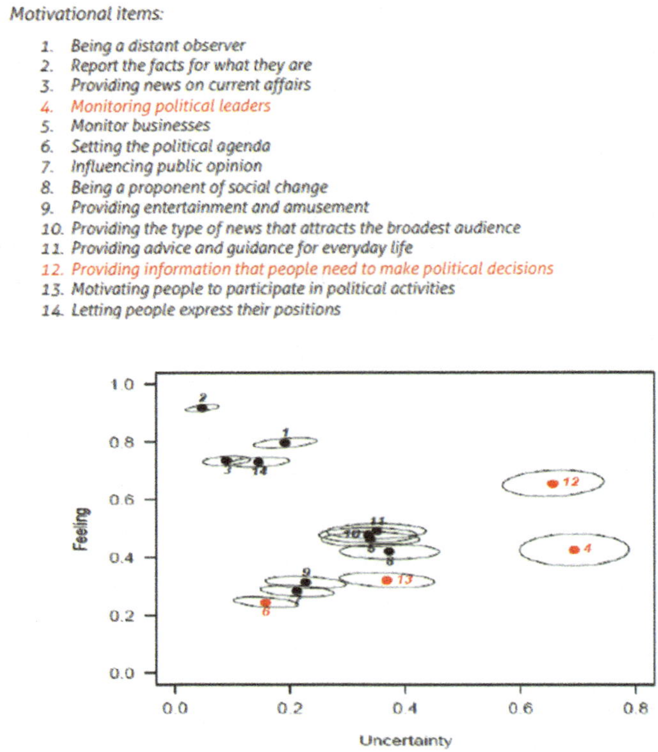

Fig. 3.1 CUB analysis of motivational items. *Source* Capecchi et al. (2020)

Based on this methodology (i.e., selection of motivational items, self-reporting questionnaire, scale purification through CUB models, identification of macro-items), two motivational indices have been constructed: i.e., the Motivational monitoring index (MMI, the maximum of items 4 and 5 of Fig. 3.1), and the Motivational provision index (MPI, item 12).

As to the **territorial context**, working in the South of Italy, for a journalist, is associated with a higher probability of suffering threats. As said. this statistical relationship, though interesting, tells us nothing about the true meaning of the link between the social phenomenon and the covariant dimension. Moreover, the choice of the granularity of the territorial data is, in the case of dichotomous variables (e.g., South = (0.1)), strictly linked to the sample size: if we move from macro-areas (i.e., North-West, North-East, Centre, South and Islands) to increasingly narrower territorial data (i.e., to regions of Italy or even counties), we obtain intersections of data with a very limited number of observations (if not even empty sets), and therefore with low or no statistical value.

The use of territorial indices solves both problems (theoretical and statistical). In the present case, what we aim to verify is whether and how much it increases, for journalists, the probability of being intimidated by the presence of criminal phenomena on the territory they work on. In this sense, two territorial indices can be employed. They are based on external sources: i.e., the Italian Institute of Statistics (ISTAT). The first, at a county level, measures the presence of common criminality (i.e., the number of crimes per 100.000 inhabitants); the second, at a regional level, measures the presence of organized crime (i.e., the number of mafia crimes per 100,000 inhabitants). By associating to each journalist, the appropriate territorial index, it is possible to conduct analyses that are both statistically finer and methodologically more focused on the phenomenon under investigation.

3.2.1.2 Composite Indicators

The move from rough data to measurement offers the advantage of framing the information according to a conceptual reference model, identifying the various dimensions that co-variate with a social phenomenon and then measuring the traits that characterize them.

The next step is the construction of composite indicators that summarize all the traits of a single dimension. As said, a composite index is the mathematical combination of a set of simple indices representing the different aspects of a dimension of a phenomenon to be measured.

The methodological steps (theoretical framework, selection of traits, multivariate preliminary analysis and managing data and statistical issues) have been summarized in the previous sections, which has also shown how different dimensions require different methodologies for the construction of single measures: the dangerousness of the activity is linked to aspects proper to journalism that imply an (iterative) assessment procedure; proximity presents objective elements of a physical nature that ease the researcher in ordering the various occurrences of different traits; motivations represent elements of a psychological nature measured by using psychometric methodologies (i.e., a self-reporting approach through a Likert scale).

The construction of a composite index addresses subsequent issues concerning the processes of normalization, weighing, and aggregation of the above indicators of simple traits.

In recent years, the increasing availability of data has stimulated the need to develop new methodologies for the construction of composite indices. In particular, Mazziotta and Pareto (2016) (see also De Muro et al. 2011) have developed an index (the so-called MPI) that is well suited for the present analysis and is increasingly used in social sciences. MPI is constructed as a combination of a "mean effect" and a "penality effect" (given by the ratio between the standard deviation and the coefficient of variation) of simple normalized indices.

By using the Mazziotta-Pareto Index (MPI), for the three dimensions analysed, for each journalist *i*, the following three composite indices were then calculated:

Dangerousness composite indicator: $DCI_i = MPI(DIT_i, DITk_i)$,
Proximity composite indicator: $PCI_i = MPI(POI_i, PAI_i, PCoI_i)$,
Motivations composite indicator: $MCI_i = MPI(MMI_i, MPI_i)$.

3.2.1.3 Data Visualization of Composite Indicators

The final stage concerns the exploratory analysis of the co-variant dimensions for the investigated phenomenon (i.e., the threats to journalists) through the newly constructed composite indices. In particular, the composite indices present some characteristics that are well suited to the exploratory data analysis (EDA) by data visualization. In fact, unlike the elementary data, they are synthetic numerical values, which measure a dimension of the phenomenon on a continuous scale (where the population average is equal to 100).

Figures 3.2, 3.3, and 3.4 illustrate the distributions of the three composite indices for two categories of journalists: those who have suffered at least one intimidation ("threatened", red-shaped distributions) and all the other journalists ("not threatened", grey-shaped distributions).

Exploratory analysis shows that all distributions are relatively more skewed to the right in the case of threatened journalists, so that their indices present an average value always higher than 100 (i.e., the average of the overall population).

This effect is mirrored, especially for the dangerousness composite index (DCI), by the existence, only for threatened journalists, of a further mode along the right tail of all the distributions (bimodal distributions).

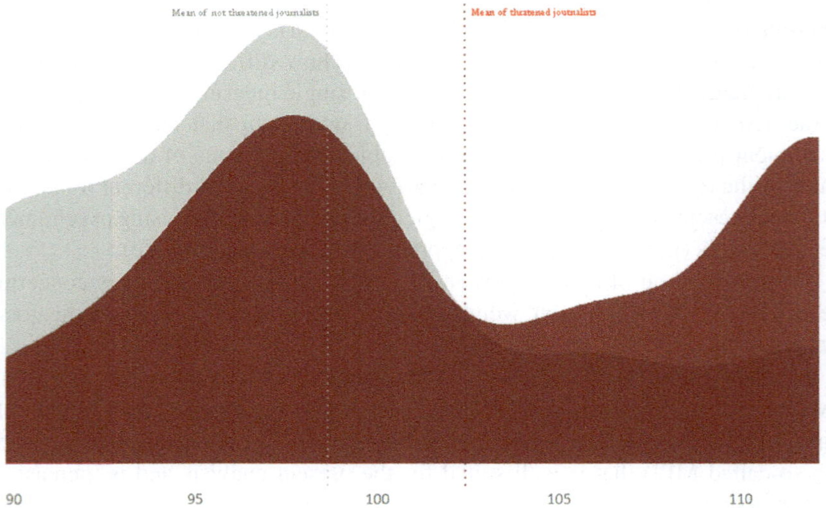

Fig. 3.2 Distribution of DCI (dangerousness composite indicator)

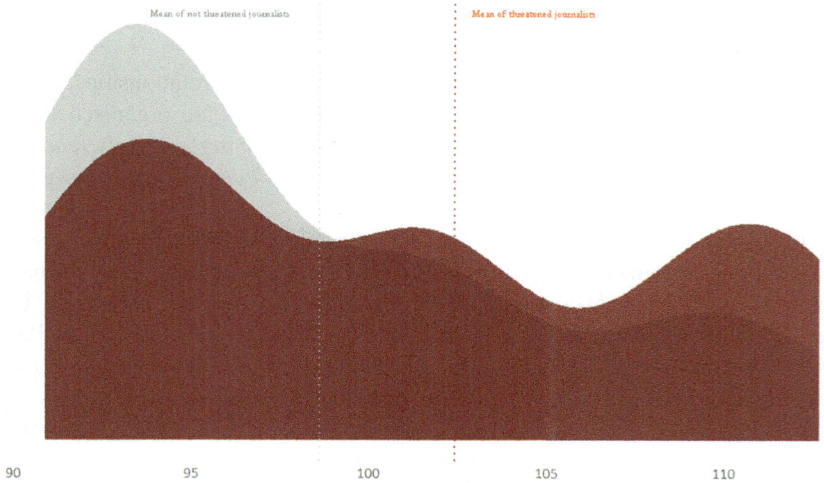

Fig. 3.3 Distribution of PCI (proximity composite indicator)

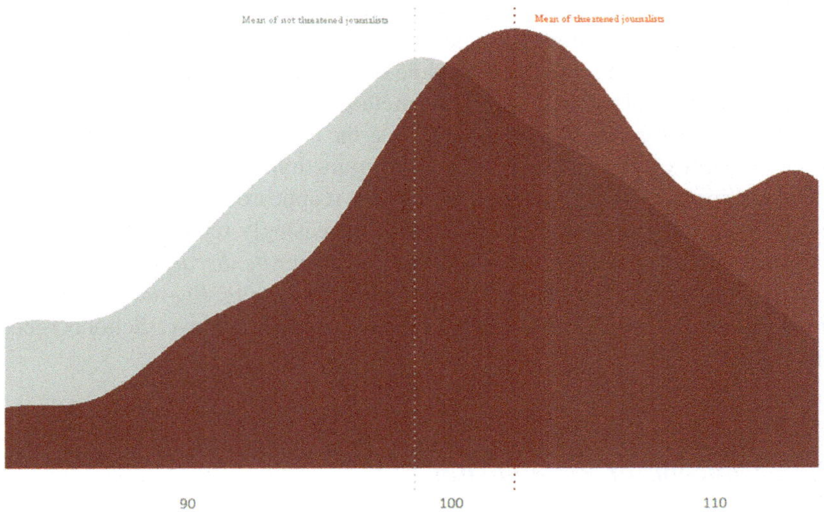

Fig. 3.4 Distribution of MCI (motivations composite indicator)

3.3 Relationships

Once the social phenomenon (through the three-stage procedure described in Chap. 2) and its covariant dimensions (as illustrated in the previous sections) have been measured, the problem moves to the relationship between them.

Here again, it is worth considering the philosophical nature of the concept of relationships between phenomena. This is a subject that has not only been debated in philosophy for several centuries (if not millennia), but also has far-reaching practical implications in terms of applied science.

Human beings are used to thinking in terms of causal relationships. Modern cognitive psychology has shown how our mental systems are organised to jump quickly to conclusions that are as causal as possible: "*We are evidently ready from birth to have impressions of causality, which do not depend on reasoning about patterns of causation*" (Kahneman 2011).

This so-called "illusion of causality" not only informs our mental processes but also drives us to pursue in science causal relationships between phenomena and events, even those that are completely improbable (what Nasssim Taleb called "black swans"). However, this may not be the case. Some philosophers, such as Pierre-Daniel Huet who wrote the *Philosophical Treatise on the Weaknesses of the Human Mind* in 1690, present arguments against causality that are quite potent: "*he states, for instance, that any event can have an infinity of possible causes*" (Taleb 2007). In essence, our predilection for causal thinking exposes us to serious mistakes in evaluating the randomness of truly random events.

But if our reference system is biased towards causality, scientific research may be equally biased towards the search for causal relationships. This might seem an idle argument, but it has fundamental scientific implications. Indeed, the Cartesian approach has had such profound scientific consequences that it has directed the theory of choice towards models of absolute rationality (see Box 3.2). In this context, causality is an integral part of the scientific method. On the contrary, modern cognitive sciences, which originate from Hume, dismiss the concept of rationality and tend to interpret causality as a constant (temporal) conjunction.

In a more practical way, the same problem has reappeared for applied science with the advent of big data, in which causality is increasingly replaced by conjunction (or more technically by correlation): "*Causality won't be discarded, but it is being knocked off its pedestal as the primary fountain of meaning. Big data turbocharges non-causal analyses, often replacing causal investigations*" (Mayer-Schonberger and Cukier 2013[27]).

3.3.1 Concepts of Relationship

To analyse the concept of relationship, it is useful to start with its negation, i.e., independence. Two phenomena are independent if the value of one provides no information about the value of the other. In probabilistic terms, if we have two variables X and Y this concept can be expressed by saying that the probability of the possible values of X is not affected by the values of Y and vice versa, i.e., $P(X|Y) = P(X)$.

---

[27] "*Rather, when we say that humans see the world through causalities, we're referring to two fundamental ways humans explain and understand the world: through quick, illusory causality; and* via*slow, methodical causal experiments. Big data will transform the roles of both*".

Association is a broad concept that also includes that of causality, in the sense that if two variables are linked by a causal relationship (e.g., X causes Y) they are also associated (i.e., they are not independent), but the opposite does not necessarily hold true. Associations are widely used in predictive models using big data. For example, our behaviour on a social network (e.g., likes) is associated with several personal characteristics that are also very sensitive (e.g., sexual orientation, political ideology up to the attitude to drug use), thus making it possible to infer, with a very high probability of success, even unconscious individual and social dimensions (Kosinski et al. 2013; Youyou et al. 2015). However, these associations tell us nothing about the causal link between the variables.

In addition, when many variables are analysed, many spurious associations are likely to arise.[28] This pattern may be due either by coincidence or by the presence of a third, unseen factor (referred to as a "confounding factor").

Technically, associations may be measured through correlations, which is a more specific concept concerning the type of relationship (trend) between two variables. This relationship can be linear (in which case the Pearson's correlation measure can be used) or non-linear (in which case it is common to use the Spearman's correlation index).

In essence, the dependence between variables is embodied in the concept of association. The association may be causal or non-causal; in the latter case the correlation measures a linear monotonic or non-linear type of association.

In any case, given that in complex systems, the number of variables is huge, this may make it practically impossible to analyse the causal relationships between the variables, and give rise to possible spurious associations. The analysis of social relations through big data not only implies a challenge in analysing causal relations, but has also highlighted how human relations are embedded in complex systems and that causation in these systems is difficult to verify, since as Huet stated, for each event there could be an infinite number of causes.

In other words, the new digital sources of data (see Chap. 2) not only impose new challenges in data management and analysis, but also force us to reconsider social science research as we finally analyse phenomena not in isolation but their natural environment, i.e., the complex social system.

3.3.2 Study Design

In Chap. 2 (see Sect. 2.3), we illustrated the importance of the study design in measuring the social phenomenon under observation, while in this chapter (Sect. 3.1) we have already highlighted the need for a study design to consider all covariant dimensions. Here, we want to emphasise that the study design may affect the type of detected relationships between variables.

---

[28] For the interested reader, I recommend Tyler Vigen's entertaining book, *Spurious Correlations* (2015), edited by Hachette Books.

Since causal relationships imply a dynamic dimensionality, surveys that are generally static are usually not suitable for measuring cause-effect relationships.[29] The variables observed through surveys are usually endogenous and it is difficult to infer the presence and direction of relationships between them. However, some factors, such as socio-demographic factors (e.g., gender and age) may be considered exogenous, i.e., they are not influenced by the other variables. For example, if one finds a positive correlation between men and smoking in a survey, we can say that being male increases the probability of smoking and we can rule out that smoking increases the probability of being male. However, we cannot say that being a man causes smoking.

At the opposite end of the spectrum of ways to collect social information are big data, which are based on unstructured data and a study design usually less controlled by the researcher. Although the data are almost always dynamic, the complexity of the system brings out more associations, even spurious ones, rather than causal relationships. As recently argued by Vitalik Buterin,[30] creator of Ethereum, the second-largest cryptocurrency by market capitalization, the internet is dominated by one-to-many and many-to-many interactions that follow strange rules that we still do not understand well: i.e., a chaotic and dense environment, difficult to model with clean and simple equations.

The desire to analyse causal relationships in the same way as in the hard sciences has led the social sciences to introduce social experiments that have become so widespread over the last 50 years that they have created their lines of research (e.g., experimental economics). In social experiments, a study design is defined such that there is a dynamic cause-and-effect relationship between two or a few factors. This is also done by identifying two groups of individuals, one undergoing treatment, the other being a control.

However, a social experiment may have several drawbacks. The first is cost, so the sample on which it is conducted is usually small and not representative of the reference population (in the past, many experiments were conducted on small samples of college students). The second is more structural and concerns the concrete possibility of isolating specific social relations. The large amount of data on social media has made us realise the complexity of social relations, and their experimental analysis appears to presuppose several strong underlying assumptions.

An interesting case in this respect is the "natural experiment". This is when individuals (or clusters of individuals) are exposed to experimental conditions that are determined by nature (e.g., a natural disaster) or by other factors outside the control of the researcher (e.g., a new normative framework). Natural experiments are widely studied in the social sciences because they potentially give the possibility to analyse social changes (and thus causal relationships) over a large group of individuals and under real conditions (not in experiments where the individual might behave very

---

[29] It is of course possible to carry out multiple surveys on the same sample of subjects to obtain dynamic data. However, this type of methodology comes up against high costs and very high implementation challenges (e.g., not all respondents to the first survey are likely to respond to subsequent surveys and this reduces the sample and may bias it).

[30] Buterin, V. (2020). The Changing Role of Economics, https://vitalik.ca/gen-eral/2020/12/28/end notes.html.

differently from real life). In this sense, the Covid-19 outbreak, which can be interpreted as a natural experiment, has already attracted a huge number of studies not only from the health sciences but also from the social sciences.

3.3.3 *Model Selection*

Model selection is the last step before the analysis of results. This should take into account many aspects of the available data, such as, among others[31]:

- Static or dynamic structure of data (e.g., survey vs. panel data model)
- Number of equations in the model (i.e., single vs. multiple equation model)
- Nature of the dependent variable(s) (e.g., bivariate, nominal, ordered, count, linear)
- Other statistical issues (e.g., sample selection bias).

3.3.4 *Threats to Journalists*

In the previous sections, we have proceeded to identify, classify and measure dimensions of social phenomena. A step-by-step procedure has been defined for the measurement of threats to journalists, where the shadow component is prevalent (see Chap. 2). Two types of measures have been defined: dichotomous, which record the possible occurrence of an intimidating event (with a differentiation according to the type of threat), and ordered categorical (or numerical), which measure the frequency of the same events (see Tables 2.1 and 2.2).

3.3.4.1 Relationship Between Threats and Covariant Dimensions

Given the nature of the metrics used to measure the dependent variable (i.e., intimidation suffered by journalists), econometric models have been estimated, respectively bivariate (for dichotomous values) and ordinal/count (for frequency values).[32]

Table 3.1 provides the results of probit models (with sample weights and robust standard errors) in which the dependent variable is the occurrence (0/1) of the following acts of intimidation over a 12-month period (see Table 2.1 for definitions): (i) 1. at least one threat, of any kind (Model I); (ii) at least one traditional threat (Model II); (iii) at least one web threat (Model III); (iv) at least one legal threat (Model IV).

---

[31] The interested reader may refer to Cox (2006).

[32] Models on frequencies (i.e., count and ordered models) are omitted for reasons of brevity.

Table 3.1 Threats to journalists: probit models

Threats	Total	Traditional	Web	Legal
Dimensions/Variables	(I)	(II)	(III)	(IV)
Role				
Journalist (D)	−0.170 (0.102)*	−0.221 (0.123)	−0.110 (0.119)	−0.226 (0.135)*
Senior Journalist (D)	0.213 (0.115)*	0.102 (0.134)	0.119 (0.130)	0.157 (0.151)
Sub-editor (D)	−0.072 (0.145)	−0.265 (0.184)	−0.202 (0.171)	0.061 (0.192)
Editor-in-chief (D)	0.323 (0.187)*	0.109 (0.234)	0.376 (0.198)*	0.662 (0.210)***
Dangerousness				
DCI (I)	0.031 (0.006)***	0.040 (0.007)***	0.022 (0.006)***	0.041 (0.008)***
Proximity				
PCI (I)	0.020 (0.007)***	0.015 (0.008)*	0.004 (0.007)	0.0193 (0.009)**
Motivations				
MCI (I)	0.026 (0.005)***	0.025 (0.006)***	0.028 (0.006)***	0.027 (0.007)***
Territory				
Mafia crimes (region, S)	0.484 (0.237)**	0.724 (0.269)***	0.466 (0.257)*	0.405 (0.280)
Total crimes (county, S)	−0.000 (0.000)	−0.000 (0.000)	–	−0.000 (0.000)
Web crimes (county, S)	–	–	0.004 (0.002)*	–
Summary stats				
Observations	1.815	1.694	1.758	1.794
Chi-squared (d.o.f.)	124.8 (11)***	103.8 (11)***	63.6 (11)***	95.1 (11)***
Pseudo R^2	0.102	0.126	0.077	0.141

Note The estimates in the table refer to probit models with sample weights and robust standard errors. The dependent variable assumes value 1 in the case of a threat in the last 12 months, 0 otherwise. Explanatory variables take the following values: "D": dummy (0/1); "I": index (means = 100); "S": statistics expressed as the number of crimes per 100,000 inhabitants. The models are estimated using the constant and socio-demographic controls (i.e., age and gender) which for reasons of space are not included in the table

For each variable, the coefficient, the standard error (in parentheses), and the level of significance are reported as follows: ***significant at 99%; **significant at 95%; *significant at 90%

As for the conjoint dimensions, as illustrated above (see Sect. 3.1.1), measures of five factors are introduced into the estimates (joint with socio-demographics—age and gender—which are omitted for space reasons): (1) role of journalists in the newsroom (baseline: freelance); (2) dangerousness of the professional activity; (3) proximity to the news; (4) professional motivations; (5) dangerousness of the working territory.

The results are statistically robust and offer interesting insights.

First, the role played by the journalist has a significant impact, especially, as expected, at the top level of the editorial pyramid: the coefficient of the dummy "editor-in-chief" is weakly significant (at 90%) in the general model (I), and strongly significant (at 99%) in the one related to legal threats (Model IV).

As for the various dimensions of professional activity analysed in the previous sections (dangerousness, proximity, and motivations), these have been introduced into the estimates using composite indicators (DCI, PCI, and MCI, respectively). As to the dangerousness of the professional activity, a strong correlation exists between the probability of suffering threats and aspects such as the topics (i.e. current affairs and politics), type of activity (i.e. investigative journalism) as mirrored by the coefficient of DCI which is positive and significant in all models.

Proximity turns out to have an overall effect on the probability of being intimidated (Model I), especially in terms of the type of activity carried out by the journalist (i.e. "field work"). As one might expect, physical proximity does not play any role when dealing with the digital ecosystem (the coefficient of PCI is not significant in the case of web threats, Model III), while it is significant both in the case of traditional intimidation and legal threats (Models II and IV).

In addition, the empirical analysis shows how motivations (especially those related to the provision of information to citizens) represent a crucial element in understanding the way journalists interpret their profession and, therefore, how much they are exposed to possible all threats from the external (criminal) environment (MCI is positive and significant in all models).

Finally, the territorial context significantly affects the probability of being threatened, in particular when the presence of mafia organized crime is higher (the coefficient of the regional index of mafia crimes is positive and significant both in the general model—Model I—and in that of traditional threats—Model II).

Figure 3.5 graphically summarizes the effect of the various dimensions of journalistic activity on the probability of being threatened. The graph shows visually the strong impact that the dangerousness of the activity and intrinsic motivations exert on the probability of suffering a threat, with this probability more than doubling from one extreme to the other of the relevant composite indices.

3.3.4.2 Assessing the Chilling Effect

The expression "chilling effect" has a very rich tradition, which can be reconstructed from an extensive scientific literature, especially in the field of legal studies on freedom of expression and opinion. This variety of themes and contributions risks rendering the concept inoperative. The phenomenon that is studied here affects the procedural aspects of freedom of expression, in practice it refers to the existence of possible factors that inhibit the exercise of free speech by the citizen, and the journalist in particular. The true essence of the chilling effect is, therefore, an act of deterrence, in relation to the activity of speaking, of exposing oneself publicly by means of one's ideas, points of view, narrations or worldviews.

This, in journalistic practice, results in the renunciation or limitation to carry out a news/investigation about a case or news event. If there are therefore possible retaliations that can be used to curb the propensity to express oneself, these are

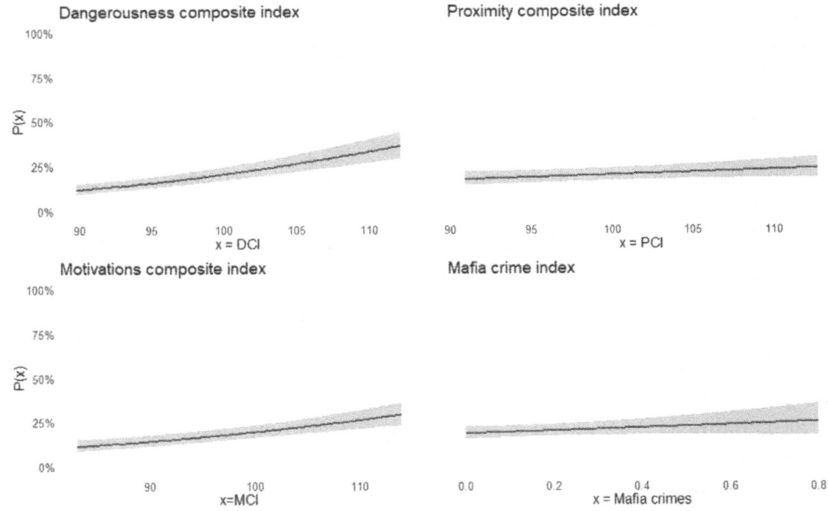

Fig. 3.5 Probability (*P(x)*) of a journalist suffering at least one threat of any kind as a function of composite indices ($x =$ *DCI, PCI, MCI, Mafia Crime Index*)

explicitly manifested through threats and intimidation of various kinds. Alternatively, deterrence may be generated by a social context that induces individuals to refrain from expressing their beliefs and opinions.

It is possible to associate to these strains two emotional levers that affect individual behaviour, i.e., the perceived risk and the fear. The perceived risk is a deterrent agent that can push individuals to prefer a state of security for themselves and for what they hold dear, rather than exercising their freedom; preserving safety always remains one of the principles at the basis of individual behaviour, and human action is in many cases an expression of this need. Fear of punishment is the basis of reluctance to perform a certain action (e.g., journalistic activity) against a written norm or social convention (as in the case of mafia contexts). In the case of the perceived risk, there is a direct link between the threats received and the complete or partial refrain from journalistic activity; in this sense, we can speak of a "direct chilling effect". In the latter case, i.e., the fear, there is instead no such direct link, and the chilling effect is of "indirect" nature, since it is generated, as said, by a broad, social context.

To translate the chilling effect into operational research, we carry out an in-depth study. In the survey (see Chap. 2), this aspect was also considered, by asking journalists whether, as a result of the intimidation received, they had reduced or even eliminated the news coverage of a topic.

From an econometric point of view, this situation, widely analysed in literature (see Heckman 1979), involves the use of a two-stage model, in which, in the first stage, the sub-sample is selected (i.e., journalists who have suffered at least one intimidation), and, in the second stage, the model estimates the relation under study (i.e., the effect of the threats on journalistic activity). This procedure allows to correct the sample-selection bias that would occur in the case of a direct analysis of the sub-sample of threatened individuals. Given the dichotomous nature of the dependent variable (which assumes a value of 1 in the case of chilling effect and 0 otherwise), a Heckman probit (weighted) model has been estimated, the results of which are presented in Table 3.2 (the selection equation has been omitted for reasons of space; see what is reported in Table 3.1, Model I).

Considering the small size of the estimated sub-sample (289 journalists) and the presence of "missing values", the results of this exercise should be taken with caution. For this reason, a limited number of covariant dimensions (i.e., role of the journalist,

Table 3.2 Chilling effect: probit two-step models

Threats	Total	Traditional
Dimensions/Variables	(I)	(II)
Role		
Free-lance (D)	0.535 (0.194)***	–
Journalist (D)	–	−0.473 (0.252)*
Senior Journalist (D)	–	−0.540 (0.255)**
Sub-editor (D)	–	−0.595 (0.522)
Editor-in-chief (D)	–	−0.878 (0.414)**
Territory		
Mafia crimes (region, S)	0.850 (0.477)*	0.914 (0.477)*
Total crimes (county, S)	– 0.107 (0.216)	-0.090 (0.217)
Type of threats		
Traditional (D)	0.512 (0.239)**	0.552 (0.238)**
Web (D)	−0.107 (0.216)	−0.090 (0.217)
ρ	−0.117 (0.392)	−0.880 (0.419)
Summary stats		
Observations	1.728	1.728
Selected observations	289	289
Chi-squared (d.o.f.)	28.9 (8)***	33.9 (11)***

Note The estimates in the table refer to Heckman probit models with sample weights and robust standard errors. The selection equation follows is omitted for reasons of space (see Table 3.1). The dependent variable assumes value 1 in the case of a chilling effect (i.e., lessening or withdrawing coverage of a topic). Explanatory variables take the following values: "D": dummy (0/1); "I": index (means = 100); "S": statistics expressed as the number of crimes per 100,000 inhabitants. The models are estimated using the constant and socio-demographic controls (i.e., age and gender) which for reasons of space are not included in the table

For each variable, the coefficient, the standard error (in parentheses), and the level of significance are reported as follows: ***significant at 99%; **significant at 95%; *significant at 90%

working territory, and type of threat received) were introduced into the estimates in addition to socio-demographic characteristics.

A preliminary comment of technical nature concerns the value of the coefficient "ρ": since it is not significantly different from zero, the null hypothesis that the two estimated equations are independent cannot be rejected (in other words there is no selection). This could be related to the fact that the two decisions (the one related to intimidation and the other related to journalistic coverage of a topic) are taken by different, and opposite subjects (who threatens and who is threatened).

First, there is a significant gender effect, with women appearing to be more vulnerable to the threats suffered (the "gender" coefficient, omitted in the table, is positive and significant in both models).

A second aspect concerns the professional status. Two factors have been distinguished in the two models shown in Table 3.2. In the first model (I) a distinction was made between freelancers and employees. As can be inferred from the coefficient of the former (positive and significant), freelances suffer more from the exposure to the threats. This appears to be linked both to a less stable and (on average) less remunerated working condition and to the fact that those who don't work structurally in an editorial office are less often protected by insurance against professional risks. In the second model (II), a distinction was made between the various roles that employees hold within the news outlet (i.e., journalist, sr. journalist, sub-editor, and editor-in-chief). The results show that as a journalist climbs up the editorial pyramid, the chilling effect is less significant.

A third facet of chilling effect concerns the type of intimidation and its "effectiveness" in shaping the information agenda. The results of the models show that "traditional" threats (i.e., warnings, assaults, and damage to personal property, see Table 2.1) display a greater relationship with the chilling effect than the other types of threats.

This latter element is linked to the last factor taken into consideration, i.e., the territorial context. Once again, it emerges how, for a journalist, working in a social environment in which mafia crimes are more frequent is structurally more critical, with a strong and significant relationship between mafia crimes and the probability of informative deterrence. The results of the models thus show the existence of a reinforcing relation between the effect of deterrence of a direct type (threat-chilling) and that of an indirect type (context-chilling). In this sense, the effect due to the perceived risk (threats) and the one connected to the fear triggered by the social context (i.e., the presence on the ground of crime organizations) tend to act in parallel, and jointly produce a much more significant impact in terms of deterrence of news production (i.e., chilling).

Figure 3.6 illustrates this last relationship, linking it to the gender effect: once suffered (at least) one threat, the probability of incurring in a chilling effect doubles if one moves from a regional context in which organized crime is absent to one in which, vice versa, it reaches its maximum levels. This relationship is much stronger for women.

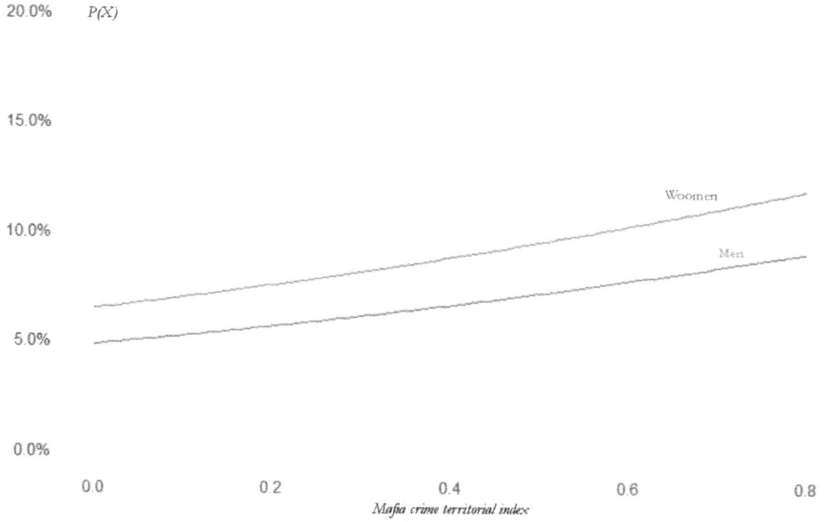

Fig. 3.6 Probability (*P(x)*) that a threatened journalist reduces or withdraws the information coverage of a topic ("chilling effect") by gender and territorial context (mafia crime territorial index)

Chapter 4
Integrating

These days, an operator takes my temperature every morning at the entrance to my office. Unfortunately, the device that measures the temperature is faulty and does not read more than 34 °C. Nevertheless, every time I tell him that this measurement is useless, he replies that it should be done anyway. Nowadays, part of science seems to me to be going the same way.

In the previous chapters, it was shown how multifaceted and multidimensional social phenomena are. This calls for an integration of data and knowledge in their analysis. However, this scientific process, which has been accelerated by the increasing availability of huge amount of data on a wide range of dimensions of individual and social life, follows an inertial and non-linear path.

The case of decision theory, mentioned in Chap. 3 (see Box 3.2), is emblematic. For centuries, science has been aware of the emotional functioning of the cognitive and decision-making systems of human beings and the fallacies of a rationalist approach (see Damasio 1994). However, it took the early work in the middle of the last century on the paradoxes of the axiomatic system related to classical decision theory to start questioning the scientific robustness of mathematical models based on expected utility. Nevertheless, arguments such as "as if" have opposed for decades the underlying criticism concerning the rationality of individuals. This is because sciences such as economics, statistics, psychology, neuroscience, and sociology have developed rather independently. Even today, cognitive psychology and neuroscience, even though they analyze (different levels of) the same domain (in neuroscience the neuronal system, while in cognitive psychology the mental systems) struggle to find a common methodological background.

But the integration of knowledge is an irreversible process, which must be enhanced by opening up the various scientific fields to different forms of expertise and methodologies. This process of openness clashes not only with the need to abandon certain scientific procedures and incorporate new ones into the analyses, but also with a change in the methodological approach. For example, at the end of Chap. 3 we showed how the availability of an unprecedented amount of data on

M. Delmastro, *On the Measurement of Social Phenomena*,
SpringerBriefs in Political Science,
https://doi.org/10.1007/978-3-030-77536-0_4

individual and collective behavior implies new methodologies of analysis and at the same time a rethinking of the Cartesian causal approach.

In a complex system such as the one in which individuals make their decisions, is it possible to isolate specific phenomena and trace their causal relationships without losing consistency? The answer is complicated and depends on the type of relationship analyzed and the goals of the analysis. However, the problem becomes more and more important as we look for robust answers upon which to base policies.

This chapter deals with the issue of integrating the social sciences. The first section illustrates some examples where integration leads to a radical change of perspective, with very relevant policy consequences. Section 4.2 introduces the topic of data integration. As these two topics are wide ranging, we explore them through two case studies that have been discussed throughout the chapters of the book, i.e., the outbreak of Covid-19 and the chilling effect due to the presence of criminal organizations.

4.1 Integrating Knowledge

In Chap. 2, we discussed the issues involved in measuring phenomena including the outbreak of Covid-19. On that occasion, we addressed the existence of problems relating primarily to measurement procedures, and therefore to official statistics, and the concrete risk of the presence of a large hidden part of the phenomenon, with consequences in terms of the effectiveness of policies based on these data and aimed at countering the spread of the virus.

In this chapter, we introduce another issue that concerns the integration of knowledge and a broader view of the epidemic. I use the example of Covid-19 to show how the integration of knowledge can broaden the scope of science and at the same time to discuss the challenges underlying a broader scientific approach.

Is Covid-19 a pandemic? Yes and no. Yes, if we observe it from an epidemiological point of view, no if we widen the scope and consider other aspects. But let us proceed in order.

First, in Chap. 2 we illustrated the various measures of phenomena such as psychological traits and in particular in Box 2.6 we outlined measurements of anxiety and depression and showed empirical evidence on the population. Using that measure (i.e., the SQMF), Giorgia Zamariola and I estimate the number of people in Italy experiencing anxiety and depression at the end of the first wave of the Covid-19 epidemic (June 2020). Note that Italy was the first western democracy to face the epidemic and its effects.

Through this methodology we can measure the relationship between the prevalence of Covid-19 and mental illness. Depression was calculated using the cut-offs of the SMRQ index that have been validated by the diagnostic literature: a score of 12 or higher may indicate the presence of clinically relevant depressive symptoms in the respondent (see Chap. 2).

Figure 4.1 shows that there is a significant relationship between the onset of depressive problems and the presence of Covid-19 in the family (Delmastro and

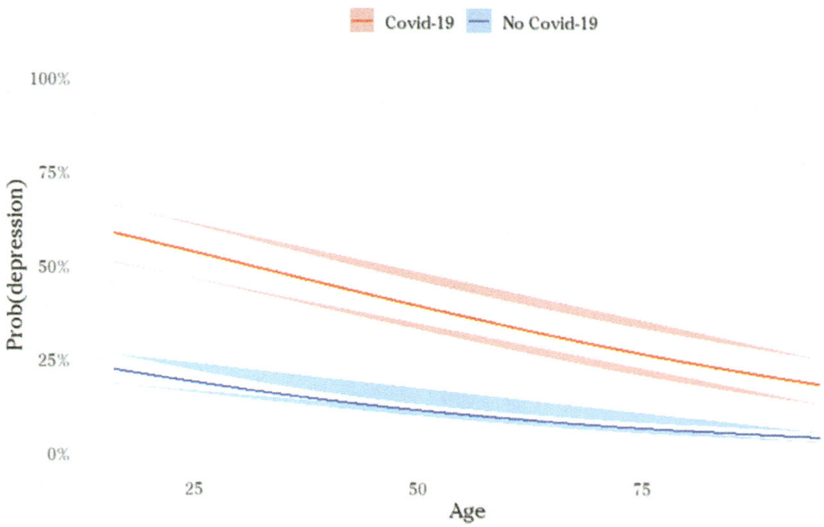

Fig. 4.1 Relationship between Covid-19 and depression in Italy. *Source* Delmastro and Zamariola (2020)

Zamariola 2020). While this relationship is very strong (more than doubling the likelihood of depressive symptoms, especially in young cohorts of the population), a causal association cannot be inferred from the data (see Sect. 3.3 for a discussion). Nevertheless, the use of psychometrics and Covid-19 diffusion measures allowed the correlation of two different pathologies to be verified. Integration has widened the scope of scientific research, which would otherwise have focused on only one topic and neglected others.

Likewise, the results of this and other research (Castriota et al. 2020) have shown that there is no correlation between the territorial spread of the epidemic and the onset of depressive symptoms. Due to the very nature of the phenomenon under investigation (i.e., an epidemic) that can potentially spread everywhere, anxiety and depression were diffused even in places where it has not yet arrived. In sum, a correlation between the spread of Covid-19 and depressive symptoms was only shown to exist where the spread directly affected a person; in the absence of this, depressive symptoms developed homogeneously throughout the country, even where the coronavirus had not yet arrived.

The second step is to test the efficacy of norms introduced to counter the spread of the COVID-19 epidemic. "Primary efficacy" of legal norms relates to the level of compliance of addressees (Grabowsky 2013),[1] where efficacy is a condition of legal validity, i.e., the norm loses its validity if it becomes permanently inefficacious: a norm "*in order to become valid, it must have the possibility of being effective*" (Kelsen 1990).

[1] "Secondary efficacy of norms" concerns the application of norms by law-applying organs.

Box 4.1 Covid-19 and disinformation

In Chap. 2, we illustrated the inherent challenges in classifying (see Box 2.2) and measuring (Box 2.5) the phenomena of mis- and dis-information. In Chap. 3, the extent to which individual choices are driven by a variety of cognitive and emotional factors is discussed (see Box 3.1). This opens up the need to explore such information consumption choices in a broad and integrated way.

To this aim, the following methods were employed which integrate socio-demographic characteristics, mental states, and cognitive abilities.

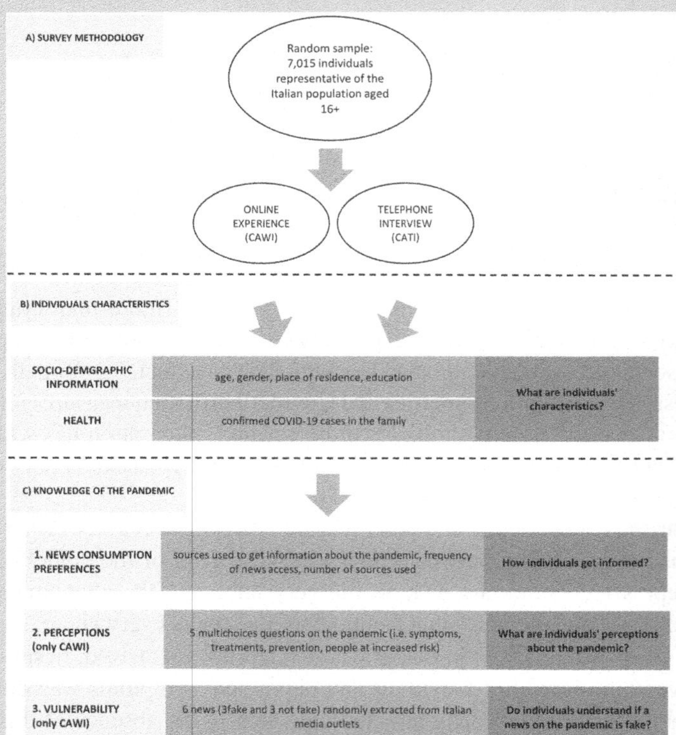

Information was gathered during the Covid-19 outbreak on a random sample representative of the Italian population (more than 7,000 individuals). Individuals were randomly presented with a set of news articles on the pandemic. Then, they were asked to indicate a level of trust (or accountability) for each news (i.e., "True/False/I don't know"). The articles consist of news that had circulated on Italian online media outlets and were randomly chosen among both fake and not fake news on the COVID-19 phenomenon.

The ability to recognize fake news from real news has been related to a variety of factors. First, in line with previous studies, less educated people result to be less able to distinguish real from fake news; this seems to be due to the circumstance that "*in the presence of motivated reasoning, education gives people better tools to counterargue against incongruent information*" (Allcott and Gentzkow 2017). We also differentiated between levels of education (i.e., primary and middle school, high school and university degrees). It comes out that only lower-level education displays a significant positive impact on the probability of errors in executive functioning in news assessment, pointing to the possible effect of functional illiteracy.

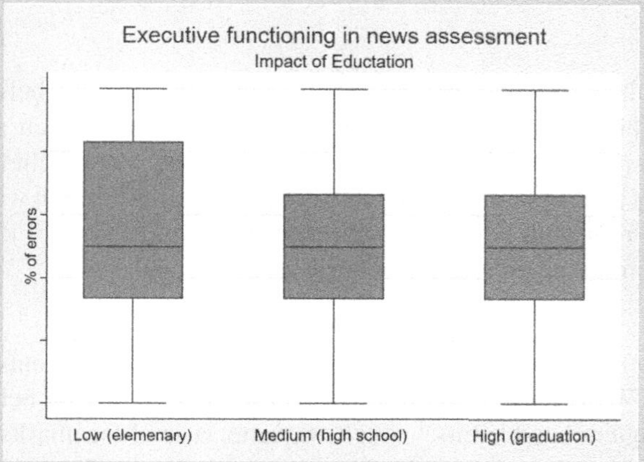

Results also indicate a strong and significant interaction between the bias in perceptions and the error probability, such that the error in news assessment is more likely for individuals that have "a priori" bias in perception about the pandemic.

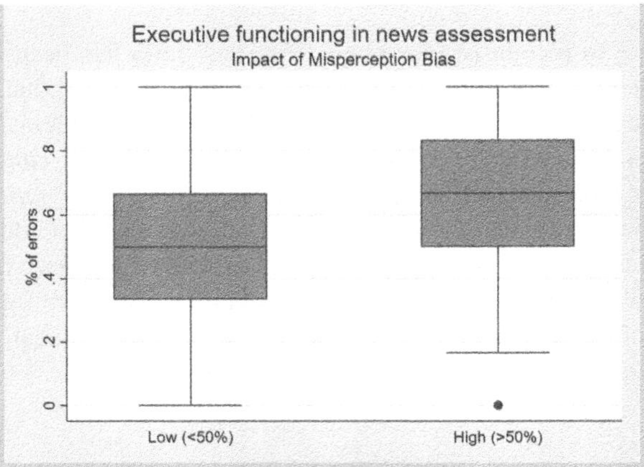

This result shows both the remarkable existence of false beliefs and the limits of scientific communication in disseminating established and verified knowledge and information to the whole population. False beliefs are at the core of many behaviors which are often considered irrational but which instead have deeper roots and which require reflection on the ability of the scientific system to communicate its knowledge (and also its failures).[2]

We consider norms that were widely employed by governments because they were based on WHO's recommendations. These provided social behavioural rules for personal hygiene[3] and habits,[4] cough etiquette,[5] use of facemasks,[6] and keeping a distance of at least one metre from persons.

In Italy, on March 9, the government signed an executive decree (i.e., "DPCM 9 Marzo 2020") that introduced these measures to the whole country, which led to a full lockdown until May 2020 (the so-called "phase 1"). Within the framework of the decree, legal rules can be summarised as follows:

- Leave home only for legitimate reasons (so-called lockdown measure).

---

[2] As suggested by the French philosopher Bruno Latour (2018) "facts remain robust only when they are supported by a common culture, by institutions that can be trusted, by a more or less decent public life, by more or less reliable media".

[3] "Perform hand hygiene frequently, particularly after contact with respiratory secretions. Hand hygiene includes either cleaning hands with soap and water or with an alcohol-based hand rub. Alcohol-based hand rubs are preferred if hands are not visibly soiled; wash hands with soap and water when they are visibly soiled".

[4] "Refrain from touching mouth and nose".

[5] "Cover your nose and mouth with a flexed elbow or paper tissue when coughing or sneezing and disposing immediately of the tissue and performing hand hygiene".

[6] "A medical mask is not required if exhibiting no symptoms, as there is no evidence that wearing a mask – of any type – protects non-sick persons. However, in some cultures, masks may be commonly worn. If masks are to be worn, it is critical to follow best practices on how to wear, remove and dispose of them and on hand hygiene after removal (see Advice on the use of masks)".

- Wash your hands often with soap and water or using an alcohol-based hand disinfectant.
- Keep a safe distance from other people (so-called social distancing).
- Cough or sneeze in the elbow crease or use a paper handkerchief, covering your nose and mouth.
- Do not touch your eyes, nose or mouth with your hands unless washed.
- Wear the facemasks according to the requirements.

These norms lasted at least until May (end of so-called "phase 1"), so that the second phase began with the end of the lockdown (May 3), the reopening of nonessential activities (May–June depending on the type of activity), and the end of the prohibition of individual movements outside people's region of domicile (June 2).

By a survey and a self-assessment methodology, we calculated a measure of compliance on a random sample representative of the Italian population (Delmastro 2020). In addition, we measured the level of compliance according to the mental state of the population, i.e., by distinguishing between people who had depressive symptoms (SQMF index equal to or greater than 12) and people who did not. Figure 4.2 shows that people with symptoms of anxiety and depression were much less likely to comply with those rules.

This second step shows that the epidemic phenomenon has endogenous components linked to the enforcement of rules.

And here we come to a more general point. Articles in the scientific literature have recently suggested that COVID-19 is something more than a pandemic (see Horton 2020; Mendenhall and Kenyon 2020). The idea is COVID-19 is a syndemic,

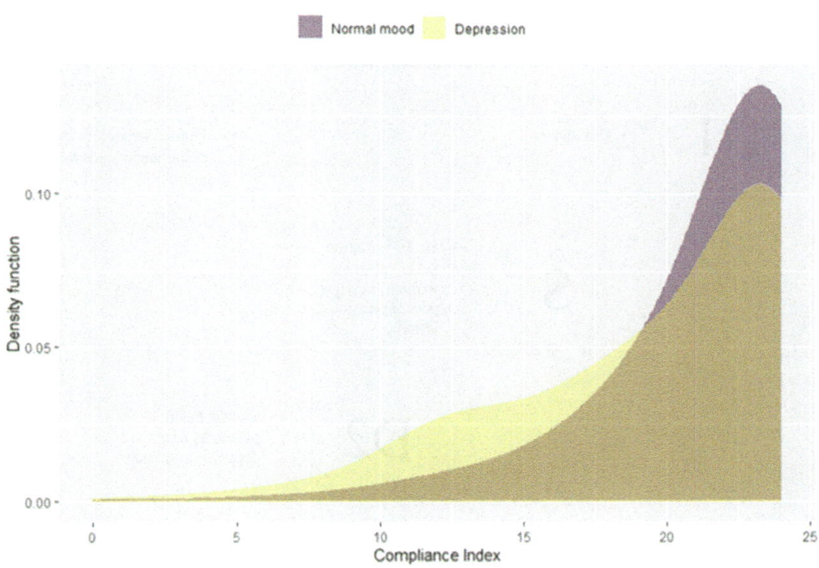

Fig. 4.2 Relationship between depression and compliance to Covid-19 norms

i.e., the interacting of sequential, or co-present, diseases and the social and environmental factors that promote and enhance the negative effects of disease interaction (Singer et al. 2017). The concept of syndemic, which originated in medical anthropology, broadens the scope of analysis not only to include the interaction between several diseases (e.g., Covid-19 and depressive symptoms, as in step 1) but also to include social and environmental factors as endogenous phenomena (e.g., institutional intervention and individual and social behaviors, as in step 2).

As shown, social compliance with health rules put in place to counter the diffusion of Covid-19 depends on mental health. Therefore, if Covid-19 is only considered as a pandemic, there is the concrete risk is to create a sort of "endogenous vicious cycle" where restrictions generate anxiety and depression, followed by a reduction in compliance and then a second contagion wave, followed by new restrictions and so on. This interactive cumulative process, which is in line with considering COVID-19 as a syndemic, is depicted in the Fig. 4.3.

In this regard, following Singer et al. (2017), Table 4.1 identifies the main factors that make this interactive process, between COVID-19 and mental health diseases in conjunction with socio-environmental factors, a syndemic.

This example shows that in many social phenomena there is a need to broaden the scope of research and integrating different knowledge. This is essential both for understanding the phenomena themselves and for developing robust policies in view of the complexity of the evolutionary processes.

Obviously, this has a price. To analyse phenomena of such complexity in a comprehensive way is very costly and requires expertise and extensive data. Moreover, the broader the scope of research, the more difficult it is to isolate causal relationships between different factors.

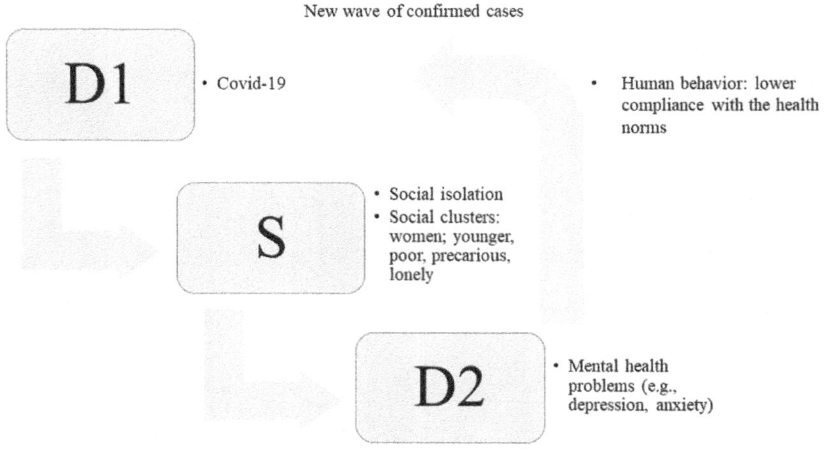

Fig. 4.3 Covid-19 as a syndemic

Table 4.1 A description of the COVID-19 syndemic in Italy

Factor	Description
Diseases (D1, D2)	D1: Covid-19 D2: Mental health diseases (e.g., anxiety, depression)
Socio-environmental conditions (S)	Social conditions influence mental health problems and interacted with Covid-19 More vulnerable social clusters: women; young people; people in conditions of poverty and professional insecurity; people living alone
Pathway (D1-S-D2)	See Fig. 4.3
Greater burden because the interaction	Increase in the number of confirmed Covid-19 cases (in the second wave) Increase number of mental health problems

4.2 Integrating Data

Data integration "*is the process of combining data from different sources into a single, unified view*" (Talend). So, integration of different data sources goes beyond the use of new sources (see Sect. 2.3). As we explained in Chap. 2, each data source may not only cover some aspects of a phenomenon, but also may suffer from statistical limitations (e.g., sample selection bias). Consequently, data integration can be aimed both at broadening the scope of the analysis and at addressing statistical issues (e.g., the need to integrate social data with census information in the case of election polls was illustrated in Chap. 2).

In this section we show the potential of data integration by bringing the case study of threats to journalists to a close. In particular, the combined use of several sources is presented. As in many cases, data integration is instrumental in covering distinct and complementary dimensions of the investigated phenomenon: in this case, on the one hand, the demand for news by citizens, on the other hand, the information supply by publishers.

4.2.1 Integrating Data for the Analysis of the News Market

4.2.1.1 Method

First, a traditional survey data will be used to monitor consumer demand for news. These data cover a large sample of the population and have been acquired by using advanced techniques (e.g., through bar codes and meters). A survey conducted in 2018 on a sample of 12,052 Italian citizens, representative of the entire Italian population over the age of 14, was then used. For each of the 20 Italian regions, individuals were selected with socio-demographic characteristics such that not only national but also regional representativeness was achieved. The sample quotas of each region were determined based on the distribution of the population by demographic size of the center of residence. Within each territorial allocation, the interviews were carried out respecting the specific quotas by gender and age class. The reference universe was determined based on official statistics available at the date of the sample definition before the interview session (i.e., ISTAT—the Italian National Institute of Statistics). When data were collected, the sample was subjected to further balancing or reweighing operations, to both take into account the deviations between the planned sample and the sample included in the survey, and to respect, in addition to the quotas assigned, the distributions according to other socio-demographic parameters. Information acquired concerns many aspects related to the news consumption of citizens, including the reading of newspapers. In addition to news consumption (in all media), information was acquired on gender, age, income, education, place of residence, profession, political orientation, and importance given to social phenomena (including the fight against the mafias).

As for the supply of news, special emphasis has been placed on general information.[7] The first step pertains to the selection of an appropriate perimeter. The aim is to ensure, on the one hand, significance in the study of the phenomenon, on the other, spatial variability. For this reason, we focus on the supply of news by local newspapers and benchmark it on that provided by national news outlets. In Italy local newspapers represent, in many areas, the most important source of news about facts. They have historically characterized the public debate since the end of the nineteenth century: *La Nazione* began publishing in 1859, *Il Giornale di Sicilia* in 1860, *La Stampa* in 1867, *Il Messaggero* in 1878, *L'Eco di Bergamo* in 1880, *Il Piccolo* in 1881, *Il Resto del Carlino* in 1885, *Il Secolo XIX* in 1886, *Il Gazzettino* in 1887, and *Il Mattino* and *La Nuova Sardegna* in 1892.

Newspapers are analysed by region according to the following classification defined by using official data (ADS and AGCOM): if the publication has a circulation equal to or greater than 75% in a county, it has been classified as "provincial" (and the newspaper has been attributed to that county/region); if the newspaper has a multi-county circulation and 75% of its circulation is within a single region, it is classified as "regional" (attributing it to the same region); if the newspaper has a circulation in several counties and more than one region, but does not reach 75% of the counties (equivalent to 83 counties), it is classified as "multiregional" (with a multi-regional attribution); finally, if the newspaper has a circulation in more than 83 counties, i.e., more than 75% of the Italian territory, it is considered as "national".

[7] Specialized news outlets (e.g., economics, finance, sport) have been excluded from the analysis.

The perimeter of general local newspapers is composed of 63 news outlets,[8] while that of national newspapers is equal to 10.[9]

Overall, the selected newspapers present characteristics of significance to the phenomenon analysed (they are the most relevant sources used by Italian citizens to inform themselves about local facts), which has deep territorial roots.

Once the whole perimeter of the newspapers has been determined, with a territorial assignment, a content analysis has been carried out. The following information is extracted from a database containing all the articles of daily newspapers in Italy for all the above mentioned local and national newspapers. The number of documents analysed is equal to 12,732,531. The measurement is of a census type, i.e., covering all articles produced by the above sources of local and national news, in the reference time horizon (i.e., 2016–2019).

We then proceed to classify the articles based on algorithmically natural language analysis and specific textual queries (see Gentzkow et al. 2019 for an introduction to the use of text as an input to social research). Textual queries refer to the presence of documents dealing with crime organizations (i.e., Camorra, Mafia, new mafias, and 'Ndrangheta) and were discussed and validated by researchers on the subject.[10]

We finally define the following metrics: (i) number of total daily documents, (ii) number of documents dealing with the topic of crime organizations; (iii) normalized documents dealing with crime organizations.

---

[8] Local newspapers analyzed (with regional assignment): *La Stampa* (Piemonte/Valle d'Aosta, Liguria), *Il Secolo XIX* (Liguria), *Brescia Oggi* (Lombardia), *Cronaca Qui* (Lombardia), *Gazzetta di Mantova* (Lombardia), *Giornale di Brescia* (Lombardia), *Il Cittadino* (Lombardia), *La Prealpina* (Lombardia), *La Provincia di Como* (Lombardia), *La Provincia di Sondrio* (Lombardia), *La Provincia di Lecco* (Lombardia), *La Provincia Pavese* (Lombardia), *La Voce di Mantova* (Lombardia), *L'Eco di Bergamo* (Lombardia), *Il Giorno* (Lombardia), *Alto Adige* (Trentino AA), *Corriere dell'Alto Adige* (Trentino AA), *Corriere del Trentino* (Trentino AA), *Il Dubbio* (Trentino AA), *L'Adige* (Trentino AA), *Trentino* (Trentino AA), *Corriere delle Alpi* (Veneto), *Corriere del Veneto* (Veneto), *Il Gazzettino* (Veneto), *Il Giornale di Vicenza* (Veneto), *Il Mattino di Padova* (Veneto), *La Nuova Venezia e Mestre* (Veneto), *La Tribuna di Treviso* (Veneto), *L'Arena* (Veneto), *Il Piccolo* (Friuli VG), *Messaggero Veneto* (Friuli VG), *Corriere di Bologna* (Emilia Romagna), *Corriere di Romagna* (Emilia Romagna), *Gazzetta di Modena* (Emilia Romagna), *Gazzetta di Parma* (Emilia Romagna), *Gazzetta di Reggio* (Emilia Romagna), *La Nuova Ferrara* (Emilia Romagna), *La Voce di Romagna* (Emilia Romagna), *Libertà* (Emilia Romagna), *Il Resto del Carlino* (Emilia Romagna), *Corriere Fiorentino* (Toscana), *Il Tirreno* (Toscana), *La Nazione* (Toscana), *Corriere dell'Umbria* (Umbria), *Corriere Adriatico* (Marche), *Il Messaggero* (Lazio, Abruzzo e Molise), *Il Tempo* (Lazio), *Il Centro* (Abruzzo), *Corriere del Mezzogiorno* (Campania), *Cronache di Caserta* (Campania), *Il Mattino* (Campania), *Il Roma* (Campania), *La Città di Salerno* (Campania), *La Gazzetta del Mezzogiorno* (Puglia, Basilicata), *Il Quotidiana di Puglia* (Puglia), *La Nuova del Sud* (Basilicata), *Gazzetta del Sud* (Calabria, Sicilia), *La Provincia di Cosenza* (Calabria), *Giornale di Sicilia* (Sicilia), *La Sicilia* (Sicilia), *Quotidiano di Sicilia* (Sicilia), *La Nuova Sardegna* (Sardegna), *L'Unione Sarda* (Sardegna).

[9] National newspapers: *Avvenire, Corriere della Sera, Il Fatto Quotidiano, Il Foglio, Il Giornale, Il Manifesto, Libero, L'Osservatore Romano, La Repubblica, La Verità.*

[10] I gratefully acknowledge Stefano D'Alfonso, Francesco Marrazzo, Rocco Mazza, and Attilio Scaglione.

4.2.1.2 How Much, When, and Where There Is Talk of Mafias

A preliminary question concerns the quantitative assessment of the observed phenomenon. In four years, local newspapers have produced over 168,000 documents (i.e., articles) about the mafias, representing an annual average of 42,008 newspaper articles. This is equivalent to 1.3% of total local news production, and to 1.9 articles per day for each newspaper.

The standardised figure was also calculated by homogenising the average daily values of articles produced by the various publications, so that newspapers containing more articles do not weigh more than newspapers with fewer news documents. The data change only marginally, with a slight increase in the number of articles produced and hence in the coverage of mafias, which rises up to 1.5% (Table 4.2).

Once the news coverage of the mafia-related topics has been defined and measured, a further analysis concerns whether and how it may vary over the week and the year. Are there days or periods of the year when this subject is most discussed in the press?

Figure 4.4 shows that Sunday and, above all, Monday are the days of the week when this topic is least covered. This is also connected to the greater space dedicated to other topics, such as sports (football in particular), which feature in the pages of newspapers on those days. However, the difference found (more than double between the minimum, 0.7% and the maximum, 1.5%) is very significant and noteworthy. Conversely, the days in the middle of the week, Wednesday in particular, are those in which it is more likely to find articles dealing with mafias' issues.

Similarly, there is a significant seasonality in news coverage (Fig. 4.5). In August the percentage of articles on the mafias is almost half of what was produced in May. The same pattern is repeated in December (another holiday month) when news coverage is again well below the average.

In short, a consolidated pattern appears to emerge according to which the news coverage of mafia events follows a regular trend, decreasing significantly in the

Table 4.2 News coverage of local newspapers of mafias (2016–2019)

Measures	Values
Articles on mafia (per year)	42,008
Normalized articles on mafia (per year)	49,510
% articles on mafias	1.3%
% normalized articles on mafias	1.5%
Articles on mafias by newspaper (per year)	697
Articles on mafias by newspaper (per day)	1.9

Note Values deriving from extractions on all articles of Italian local newspapers, in the period 2016–2019. The "normalized" values are calculated to homogenize the different average number of daily articles in the various newspapers

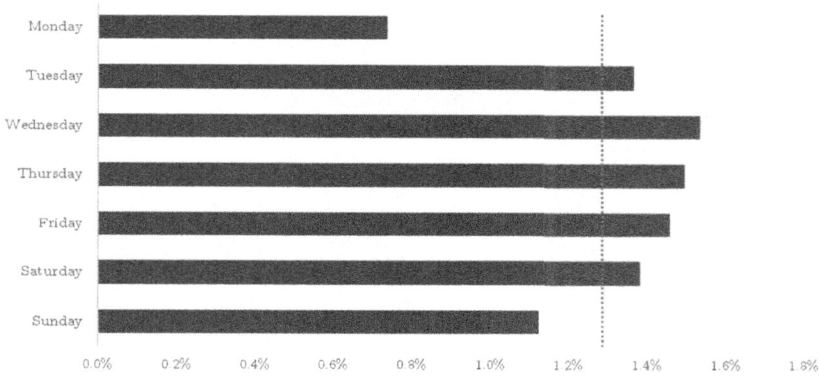

Fig. 4.4 Daily coverage of local newspapers of mafias (2016–2019). *Note* % articles on mafias from local Italian newspapers, during 2016–2019

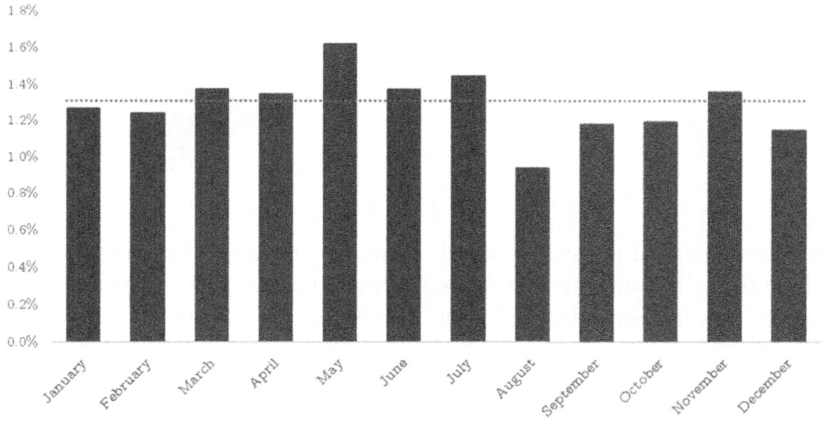

Fig. 4.5 Monthly coverage of local newspapers of mafias (2016–2019).[11] *Note* % articles on mafias from local Italian newspapers, during 2016–2019

moments of the week and of the year, in which the news is dedicated to light topics, and therefore, there is less space for news on organized crime.

Mafias were born as socio-criminal phenomena of local nature connected to some areas of Italy, essentially, in Sicily "Cosa Nostra", in Campania the "Camorra", in Calabria the "'Ndrangheta" and in Puglia the "Sacra corona unita". They represent, therefore, a social element, territorially rooted.

In the light of these considerations, the news coverage should show a significant geographical differentiation. This circumstance is confirmed by the analysis of

[11] This result reminds me of a quote from a famous Italian film: «*Do not worry. Now it's winter. The mafia only kills in summer*» (Arturo's Dad, in *La Mafia Uccide solo d'Estate* (Mafia Kills Only in Summer).

Fig. 4.6 Territorial coverage of local newspapers of mafias (2016–2019). *Note* Regional distribution of the % of mafia-related articles. In dark blue, the regions with news coverage on mafias greater than 3.5%. In blue, the regions with news coverage on mafias between 3.5% and 2%. In light blue, the regions with news coverage on mafias lower than 2%

the data, summarized in Fig. 4.6, which shows how the aforementioned regions of Southern Italy (in particular, Campania and Sicilia) present rates of news coverage of mafia events far more significant than elsewhere.

In addition to the aspects of local roots, there are other elements, always linked to mafia, which concern more general considerations of a national (or even international) nature. Criminal organizations, over time, have gradually expanded throughout the Italian territory (and abroad), reaching economic interests outside their traditional regions, especially in the northern richer part of Italy. Furthermore, there has been the emergence of new mafias, some of an endogenous nature (see the case of Rome), others of an exogenous nature (for example, the so-called "Nigerian mafia"). These aspects will be considered in the following analysis.

4.2.1.3 Is There Enough News Coverage of Mafia? The Chilling-Effect Revisited

In the previous Chapter (see par. 3.3.4b) it has been shown that there is a deterrent effect on news (i.e., chilling effect), resulting from the numerous intimidating episodes suffered by journalists. This was done through the study of a sample of journalists and their reactions, in terms of limiting information coverage, to any threats received. In that study, some determinants were tested. Among these, the role of the territorial context emerged tied to the presence of organized crime of mafia type.

In this paragraph, after having quantified the phenomenon and some of its traits, we analyse the same problem from a different point of view, namely the overall supply of news. The issue is very complex and concerns the existence (or not) of a gap between how much information is produced and how much would have been offered in the absence of deterrent factors.

Considering that the current supply of news content must be compared with a hypothetical (therefore non-existent) level, three possible methodological methods are proposed below, none of which should be considered definitive. Nevertheless, each of them offers an interesting methodological clue, complementary in providing an overall explanation to the research question.

The first method is a standard benchmark analysis. The feasibility of making a methodologically correct comparison is not easy in this case. However, the available data can be used to acquire some limited but significant evidence.

In particular, it has been chosen to compare the supply of news of local newspapers with that of national newspapers. This makes it possible to compare news outlets on an equal footing, in terms of: (i) textual and semantic patterns of reporting mafias; (ii) object of analysis (i.e., articles), and (iii) structure of the service provided to the public (paid for news in both printed and digital form).

The metric used for the benchmarking is the percentage of articles on mafias out of the total number of articles published. Since there are some differences in the collection of the two types (national and local) of newspapers,[12] the coverage ratio of the topic appears to be the most appropriate measure to operate a benchmark analysis.

Table 4.3 reports this measure (and the relative difference) for both types of newspapers, and for three groups of them that have been identified from a territorial point of view: (i) all newspapers (local and national); (ii) only southern newspapers (or at least centre-south newspapers); (iii) only newspapers (national and local) based in Rome.

The aim is to benchmark groups of newspapers that are increasingly homogeneous from a territorial point of view, although presenting a different information profile: national vs. local. Since the national newspapers are based either in Rome or Milan, the only possible comparison on a specific local focus affected by mafias is the one related to the Italian capital city (and its county).

Figures show that, on average, the rate of coverage of local newspapers on mafias is, with similar territorial footing, in all cases lower than of the national publications; this difference is significant (by statistical tests) and it is about 1 percentage point

[12] Some national newspapers offer a part dedicated to local area that is not present in the gathering of them. This leads, for these newspapers, to a smaller number of articles produced per newspaper; but this value is a spurious one because the result of the different collection, not the effect of a different informative offer.

Table 4.3 "Local vs. National": differences in news coverage on mafias (2016–2019)

Newspapers: Local–National	% articles on mafias	Difference (%)
All	1.3–2.7	−1.4
Southern	2.9–3.6	−0.7
Rome-based	2.3–3.5	−1.2

Note Values deriving from local newspapers and national newspapers in the period 2016–2019. The "Southern local newspapers" refer to those attributable to the regions of the Centre-South of Italy (20 out of the 63 analysed as a whole). "Southern national newspapers" are those based in the Centre-South of Italy (i.e., in Rome, 3 out of 10, because there are no national newspapers south of Rome). Articles in the national newspapers do not cover the local news sections

(between 0.7% and 1.4%), i.e., between a third and a half of all the news production about mafias.

Such evidence might suggest that the proximity of the news outlet to a territory with a high presence of criminal organizations is more likely to make it more exposed to chilling effect phenomena, not just of a direct type, see Table 3.2 in Chap. 3 in the case of journalists, but even of an indirect type. In other words, a local newspaper would experience more difficulties, especially financial ones, in sustaining possible (legal) retaliations by criminal lobbies and, therefore, would tend to choose a generally more precautionary editorial line compared to a national newspaper.

Some methodological limitations need to be taken into account. The previous comparison is, from a territorial point of view, completely identified only for the (national and local) newspapers based in Rome, which represent an important but very limited case. Moreover, the differences found could be due to other editorial reasons, although most of them would lead to believe that the coverage of mafia should be greater in a local newspaper than in a national publication; for instance, national newspapers cover subjects, such as foreign affairs, which are generally not included in local papers.

Even more important is the question of how much the comparison with the national coverage can represent an efficient benchmark. After all, as depicted in Fig. 4.7, even national newspapers can be subject to a chilling effect and, therefore, at most, they may provide a minimum threshold for the supply of content on mafia, not a frontier of efficiency.

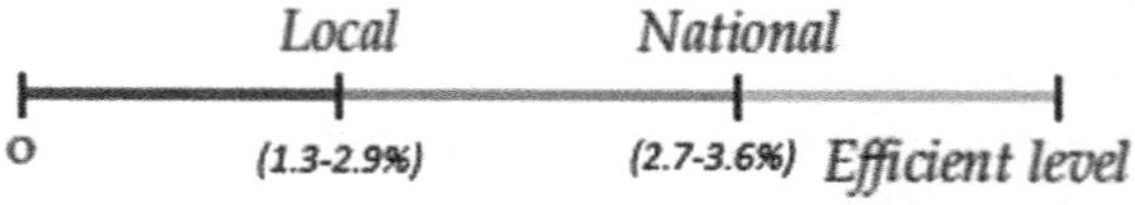

Fig. 4.7 Benchmark analysis and efficient level of news coverage of mafias

If the national news offer does not constitute an efficiency benchmark, but only a reference threshold, the next question is how to estimate an efficient level, i.e., not conditioned by the deterrent effect, given that the latter, as shown in the previous Chapter is at work.

All other things being equal, the relationship between mafia events and news coverage should be increasing. A positive relation holds for every significant event, i.e., for every event that presents characteristics of newsworthiness. There is a sort of production function that ties the events to the news published. This relationship can be linear, when the number of news documents on a topic grows in a constant proportion to the events related to the topic itself. When the number of news documents grows at decreasing rates, then the function is concave. This latter pattern could be determined by a saturation phenomenon, meaning that as the number of events increases the novelty and the newsworthiness of the theme tend to decrease. However, since we are considering a long-time span (four years), not a point in time, this effect may be less relevant; on the contrary, at the beginning of the relationship the opposite effect (convexity) may hold true: a series of events can induce newsworthiness of a topic, which in turn can determine an initial news coverage more than proportional to the events themselves.

If we introduce the deterrent phenomenon (i.e., the chilling effect) generated by a criminal environment, this leads the news function to be concave (eventually flat): as the number of mafia crimes increases, the direct (perceived risk due to threats) and indirect (general context of fear) pressure on journalists and news outlets increases, thus reducing news coverage of crime events.

In addition to the aspects of close local roots, there are, as said, other elements, always linked to mafia, which concern more general considerations of a national (or even international) nature. Criminal organizations, over time, have gradually expanded throughout the Italian territory (and abroad), reaching economic interests outside their traditional regions, especially in the northern richer part of Italy.

In sum, there are two informative components on the subject of mafias. The first locally grounded, related to crimes committed on the territory, the second of a national nature, connected to other socio-economic aspects. To acquire some knowledge of both components, the regional average news coverage was analysed according to the number of mafia-type crimes committed on the regional territory. The latter have been measured through official statistics which, on a territorial basis, report, every year, two types of mafia crimes: "voluntary mafia-type homicides" and "conspiracy to commit mafia-type crimes".

In Fig. 4.8 the estimated relationship between mafia crimes and news coverage is depicted, including (in addition to the interpolant) also those functions that pass through the maximum and minimum effective level of news coverage. Finally, the linear line represents an ideal benchmark in case the proportion (i.e., the slope of the news production function) between news and events holds constant.

Both components of news coverage can be read from the graph. The first national component determines the y-intercept of the interpolating function: in the absence of mafia crimes committed on the territory (mafia crimes = 0), there is, in any case, a news coverage (around 0.5–0.8%) connected to elements of a general and

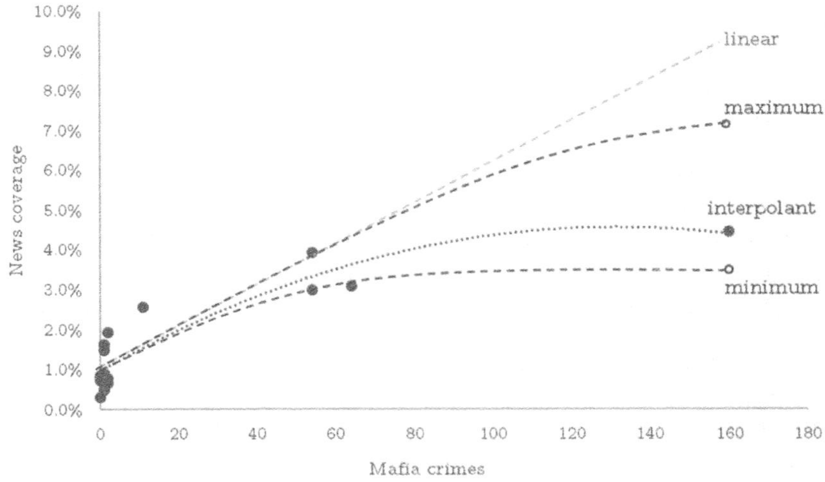

Fig. 4.8 Estimated relationship between mafia crimes and news coverage. *Note* The crimes considered are "mafia-type voluntary homicides" and "conspiracy to commit mafia-type crimes". *Source* ISTAT; number of crimes reported by the police to the judicial authorities

national interest in the mafia phenomenon. The second characterizes the relationship (of a logarithmic type) between crimes committed on the territory by criminal organizations and the news coverage in the region.[13] As mafia crimes grow, the difference between the interpolant and the linear function, or in any case concerning the maximum effective value, can be reasonably interpreted as the result of the working (also) of the chilling effect, whose presence has already been detected (at micro level) in the previous Chapter (see Sect. 3.3.4.b).

So far, the side of news production has been analysed, both from the point of view of journalists (in Chap. 3) and news outlets (in this section). The last step is to check how much supply and demand meet.

Although it may seem obvious that in a market supply and demand will balance in equilibrium, in the case of the "news market" this may not be necessarily true (see Gentzkow and Shapiro 2008). First, most of the information is free (think of public broadcasting, commercial radio and TV outlets, and many news websites). In this case, no price mechanism balances supply and demand. In addition, even in the case of paid for news content (as in the case of newspapers), the consumer purchases a comprehensive bundle of local, national and international news. In this context, consumers' demand for news on specific issues (such as mafias) may not find an adequate supply-side counterpart.

The analysis of consumers' purchasing decisions is therefore crucial to understand how well the news market manages to bring together supply and demand, especially

[13] A stochastic frontier model has been regressed. Results show that the relationship between news coverage and crimes is positive, significant (at 99%) and logarithmic (the linear coefficient is instead insignificant).

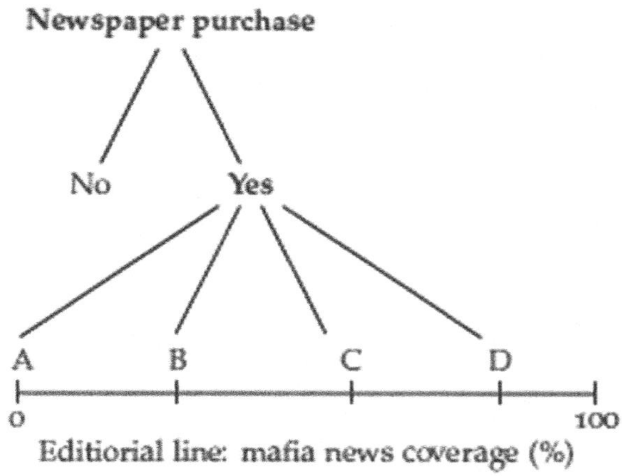

Fig. 4.9 Model of newspapers' consumption: two-stage model

on selected editorial content. To analyse the matching between supply and demand of information content on mafia, we proceed, as said, by integrating different sources, as follows.

First, all the Italian national newspapers were considered according to the methodology illustrated above. These newspapers were then distributed along a horizontal axis according to the coverage rate attributed by them to mafia events in the period 2016–2017. Second, a survey conducted in 2018 on a sample of 12,052 Italian citizens, representative of the entire Italian population over the age of 14, was then used (see above). Finally, the individual choice to purchase a newspaper was modelled as follows.

The decision-making process has been divided into two logically distinct and sequential steps (see Fig. 4.9). In the first step, the consumer decides whether or not to buy a newspaper. This decision is widely discussed in the literature, also with regard to possible forms of complementarity with the consumption of other media (see Gentzkow 2007). In the second step, the consumer decides which newspaper to buy based on the editorial line followed by the newspaper (i.e., in the period before the survey, 2016–2017). In this case, the editorial line has been measured by the news coverage rate of the mafia-related events, and can be represented as a point on a horizontal line ranging from 0 (no coverage) to 100% (full coverage). News coverage of mafias in national newspapers, in the period 2016–2017, ranges from a minimum of 0.8% to a maximum of 9.1%, showing a considerable variability between newspapers.

This approach is consistent with the Lancaster's view that goods can be summarised by a multidimensional bundle of intrinsic characteristics, on which consumers make their own choices. In the case of Italian newspapers, since price is not a variable that differentiates the various titles (due to a historical homogeneity of price strategies, see Argentesi and Filistrucchi 2007), the competition has traditionally focused on the editorial line (so-called horizontal differentiation, à la Hotelling).

A two-stage model was then estimated: in the first stage (selection equation) the decision whether or not to buy a national newspaper is regressed, while in the second stage the decision on the specific newspaper is estimated according to the editorial line preferred, among those existing, by the consumer. The results of the estimates are presented in Table 4.4.

A preliminary consideration concerns the value of the "ρ" coefficient which, in this case (unlike the model on the chilling effect, see Table 3.2 in Chap. 3, where the two equations could be considered independent from one another), is significant. This means that the two decisions (i.e., purchase of the newspaper and choice of the newspaper according to the editorial line) cannot be considered independent one from the other and must therefore be considered, as we would have expected, in conjunction.

As regards the purchase decision (i.e., the selection equation whose results are shown at the bottom of Table 4.4), the coefficients of the explanatory variables are those expected, in line with international evidence (Gentzkow 2007). The decision to buy (or not) a newspaper is influenced by the gender (newspapers are mostly read by a male audience), age (older population reads more newspapers), education (the coefficient of graduation is positive and significant), economic condition (newspapers are read mostly by wealthy individuals) and place of residence (newspapers are read more in big cities). In addition, there appears to be a complementary relationship with other media, especially with magazines (weekly and monthly) and the internet.

The selection of news outlets based on the editorial line shows interesting and innovative results. First, many variables which are significant at the purchasing stage lose importance in the model regarding the specific choice of the newspaper. For example, income and occupational variables display no effect, since in Italy, as mentioned above, there is considerable price homogeneity; consequently, the diversification among the socio-economic groups takes place in the first step, not afterwards.

The choice of the newspaper according to the editorial line is substantially driven by three personal traits: (i) where one lives, (ii) what ideological orientation one has, and (iii) what importance one assigns to the topic.

It is interesting to notice that the editorial line is an intrinsic trait, not a well-known commercial brand. However, individuals seem to orient themselves between the editorial lines of the different national titles.

Citizens living in the southern parts of Italy, as seen, with a higher intensity of mafia crimes, turn to newspapers that cover this issue more. This is true, in strict analogy, also for those who live in large cities (not only in the South), where the social effects of the organized crime are more harmful, and, in any case, more perceived.

The political orientation of citizens, and therefore their point of view in the public debate, directs the information choice: given Italian political fragmentation, four political orientations have been distinguished (i.e., X, Y, W and Z) and confronted with those who claim not to have one, which has been chosen as the baseline of the estimates. This is a well-known result in media literature, but it is interesting that it emerges also for a specific editorial topic, such as that of mafia crimes.

Finally, all other things being equal, those who believe that the fight against mafia should be a key element of the national policy (i.e., 21.6% of Italians in 2018), tend

Table 4.4 Newspaper choice: Hechkman two-step model

Variables	Newspaper choice (editorial line)
Gender (woman, D)	0.091.(0.080)
Age (years, A)	0.003 (0.004)
Education (graduation, D)	0.098 (0.111)
North-west (D)	−0.469 (0.067)***
Centre (D)	0.157 (0.105)
South-Islands (D)	0.590 (0.114)***
City (more than 500,000 inhabitants, D)	0.255 (0.113)**
Professional (D)	−0.229 (0.135)*
Shopkeeper (D)	0.181 (0.191)
Executive (D)	−0.045 (0.154)
Clerk (D)	−0.089 (0.117)
Housemaker (D)	−219 (0.142)
Student (D)	0.094 (0.211)
Retiree (D)	−0.1230 (0.142)
Unemployed (D)	−0.212 (0.123)*
Political orientation X (D)	−0.180 (0.073)**
Political orientation Y (D)	0.093 (0.092)
Political orientation W (D)	−0.166 (0.103)
Political orientation Z (D)	0.087 (0.083)
Importance of policies fighting mafias (D)	0.199 (0.086)**
Variables	Newspaper buying (selection)
Gender (woman, D)	−0.444 (0.042)***
Age (years, A)	0.012 (0.002)***
Education (graduation, D)	0.340 (0.056)***
Income (C)	0.078 (0.019)***
North-west (D)	0.292 (0.058)***
Centre (D)	0.139 (0.065)**
South-Islands (D)	−0.098 (0.058)*
City (more than 500,000 inhabitants, D)	0.186(0.056)**
Professional (D)	0.548 (0.089)***
Shopkeeper (D)	0.247 (0.089)***
Executive (D)	0.259 (0.113)**
Clerk (D)	0.318 (0.068)***
Housemaker (D)	0.035 (0.089)

(continued)

Table 4.4 (continued)

Variables	Newspaper buying (selection)
Student (D)	0.455 (0.102)***
Retiree (D)	0.099 (0.088)
Unemployed (D)	0.269 (0.078)***
Pay TV (subscription, D)	0.034 (0.045)
Radio (all days, D)	−0.014 (0.077)
Weekly magazine (D)	0.651 (0.041)***
Monthly magazine (D)	0.407 (0.041)***
Internet (all days, D)	0.142 (0.043)***
ρ	−0.139 (0.049)***
Observations	12,052
Selected observations	2,903
Chi-squared (d.o.f.)	192.65 (20)***

Note The estimates in the table refer to Heckman probit models with sample weights and robust standard errors. Explanatory variables take the following values: "A" number of years; "C": categorical; "D". dummy (0/1). For each variable, the coefficient, the standard error (in parentheses), and the level of significance are reported as follows: ***significant at 99%; **significant at 95%; *significant at 90%

to select a newspaper that embraces an editorial line more oriented to deal with this issue.

In summary, the results of the model show that, despite the absence of an efficient market mechanism (due to the specific characteristics of the "news market"), there appears to be a significant sensitivity of demand to the editorial line of newspapers, even when we focus on a particular topic such as mafias.

Selected References

Allahyari M, Pouriyeh S, Assefi M, Safaei S, Trippe E, Gutierrez J, Kochut K (2017) A brief survey of text mining: classification, clustering and extraction techniques. arXiv e-prints
Argentesi E, Filistrucchi L (2007) Estimating market power in a two-sided market: The case of newspapers. J Appl Economet 22:1247–1266
Bailey K (1994) Typologies and taxonomies: an introduction to classification techniques. Sage, Thousand Oaks
Blazquez D, Domenech J (2018) Big Data sources and methods for social and economic analyses. Technol Forecast Soc Chang 130:99–113
Bohrnstedt GW (2010) An overview of measurement in the social sciences. In: Paper prepared for the Workshop on advancing social science theory: the importance of common metrics. National Academies, Washington, DC, 25–26
Borsboom D (2005) Measuring the mind: Conceptual issues in contemporary psychometrics. Cambridge University Press, Cambridge
Bovet A, Morone F, Makse H (2018) Validation of Twitter opinion trends with national polling aggregates: Hillary Clinton vs Donald Trump. Sci Rep 8:8673. https://doi.org/10.1038/s41598-018-26951-y
Butler D (2013) When google got Flu Wrong. Nature 494:155–156
Campbell N (1957) Foundations of Science: the philosophy of theory and experiment. Dover Publications, New York
Cappella J (2017) Vectors into the future of mass and interpersonal communication research: big data social media, and computational social science. Hum Commun Res. https://doi.org/10.1111/hcre.12114
Cartwright N, Ruhnhardt R (2014) Measurement. In: Cartwright N, Montuschi D (eds) Philosophy of science: a new introduction. Oxford University Press, Oxford
Cartwright N, Bradburn N (2016) A theory of measurement. CHESS Working papers
Castriota S, Delmastro M (2015) The economics of collective reputation: evidence from the wine industry. Am J Agr Econ 97:469–489
Castriota S, Delmastro M, Tonin M (2020) National or local? The demand for news in Italy during Covid-19. CESifo Working Paper, 8699
Chang H (2009) Operationalism. In: Zalta E (ed) Stanford Encyclopedia of Philosophy
Colombo M, Delmastro M (2008) The economics of organizational design: theoretical insights and empirical evidence. Palgrave Macmillan, London
Coltman T, Devinney T, Midgley D, Venaik S (2008) Formative versus reflective measurement models: two applications of formative measurement. J Bus Res 61:1250–1262

M. Delmastro, *On the Measurement of Social Phenomena*,
SpringerBriefs in Political Science,
https://doi.org/10.1007/978-3-030-77536-0

Cortina J (1993) What is coefficient alpha? An examination of theory and applications. J Appl Psychol 78:98–104
Cox D (2006) Principles of statistical inference. Cambridge University Press, Cambridge
Cox D, Kartsonaki C, Keogh R (2018) Big data: some statistical issues. Statist Probab Lett 136:111–115
Damasio A (1994) Descartes' error: emotion, reason, and the human brain. Putnam Adult
Delmastro M (2020) To conform or not to conform? Legal norms and human behaviour during the COVID-19 pandemic. Research Square. https://doi.org/10.21203/rs.3.rs-127156/v1
Delmastro M, Splendore S (2020) Google, Facebook and what else? Measuring the hybridity of Italian journalists by their use of sources. Eur J Commun 17. https://doi.org/10.1177/0267323120940912
Delmastro M, Zamariola G (2020) Depressive symptoms in response to COVID-19 and lockdown: a cross-sectional study on the Italian population. Sci Rep 10:22457. https://doi.org/10.1038/s41598-020-79850-6
De Muro P, Mazziotta M, Pareto A (2011) Composite indices of development and poverty. Soc Indic Res 104:1–18
Di Nardo M et al (2005) Handbook on constructing composite indicators. OECD Statistical Working Papers
Edelmann A, Wolff T, Montagne D, Bail C (2020) Computational social science and sociology. Ann Rev Sociol 46:61–81
Gentzkow M (2007) Valuing new goods in a model with complementarity: online newspapers. Am Econ J 97:713–744
Gentzkow M, Kelly B, Taddy M (2019) Text as data. J Econ Lit 57:535–574
Gentzkow M, Shapiro J (2008) Competition and truth in the market for news. J Econ Perspect 22:133–154
Grčar M, Cherepnalkoski D, Mozetič I, Kralj Novak P (2017) Stance and influence of Twitter users regarding the Brexit referendum. Comput Soc Networks 4. https://doi.org/10.1186/s40649-017-0042-6
Halford S, Savage, M (2017) Speaking sociologically with big data: symphonic social science and the future for big data research. Sociology 51:1132–1148
Hauser J, Katz G (1998) Metrics: you are what you measure! Eur Manag J 16:517–528
Hayes E (1911) Classification of social phenomena. Am J Sociol 17:90–1118.
Heckman J (1979) Sample selection bias as a specification error. Econometrica 47:153–161
Heilmann C (2015) A new interpretation of the representational theory of measurement. Philos Sci 82:787–797
Horton R (2020) Offline: COVID-19 is not a pandemic. Lancet 396:874
Jarvis CB, Mackenzie SB, Podsakoff PM (2003) A critical review of construct indicators and measurement model misspecification in marketing and consumer research. J Consum Res 30:199–218
Jordi C (2021) Otto Neurath. In: Zalta E (ed) The Stanford Encyclopedia of Philosophy
Kahneman D (2011) Thinking, fast and slow. Penguin
Kosinski M, Stillwell D, Graepel T (2013) Private traits and attributes are predictable from digital records of human behavior. PNAS 110:5802–5805
Krantz D, Luce R, Suppes P, Tversky A (1971) Foundations of measurement Vol 1: additive and polynomial representations. Academic Press, London
Latour B (2018) Down to earth: politics in the new climatic regime. Polity, London
Liu H, Motoda H (2012) Feature selection for knowledge discovery and data mining. Springer Science, New York
Mari L (2005) The problem of foundations of measurement. Measurement 38:259–266
Mayer-Schonberger V, Cukier K (2013) Big data: a revolution that will transform how we live, work, and think. Houghton Mifflin, Harcourt, Boston
Mazziotta M, Pareto A (2016) On a generalized non-compensatory composite index for measuring socio-economic phenomena. Soc Indic Res 127:983–1003

Mendenhall E (2020) The COVID-19 syndemic is not global: context matters. Lancet 396:1731
National Research Council (2011) The importance of common metrics for advancing social science theory and research: a workshop summary. The National Academies Press, Washington, DC
OECD (2008) Handbook on constructing composite indicators: methodology and user guide. OECD Publishing
Schintler L et al (eds) (2021) Encyclopedia of big data. Springer Nature, Cham
Simon H (1972) Theories of bounded rationality. In: McGuire C, Radner R (eds) Decision and organization. North-Holland Publishing, Amsterdam
Singer M, Bulled N, Ostrach B, Mendenhall E (2017) Syndemics and the biosocial conception of health. Lancet 389:941–950
Stevens S (1951) Mathematics, measurement, psychophysics. In: Stevens S (ed) Handbook of experimental psychology. Wiley, New York, pp 1–49
Szostak R (2004) Classifying science. Springer, Dordrecht
Tal E (2013) Old and new problems in philosophy of measurement. Philos Compass 8:1159–1173
Tal E (2020) Measurement in science. In: Zalta E (ed) Stanford Encyclopedia of Philosophy
Taleb N (2007) The black swan: the impact of the highly improbable. Random House and Penguin
Thabrew H, Stasiak K, Bavin LM, Frampton C, Merry S (2018) Validation of the Mood and Feelings Questionnaire (MFQ) and Short Mood and Feelings Questionnaire (SMFQ) in New Zealand help-seeking adolescents. Int J Methods Psychiatr Res 27:1–9
Van Bavel J, Baicker K, Boggio, PS et al (2020) Using social and behavioural science to support COVID-19 pandemic response. Nat Hum Behav 4:460–471
Varian H (2014) Big data: new tricks for econometrics. J Econ Perspect 28:3–28
Vessonen E (2017) Psychometrics versus representational theory of measurement. Philos Soc Sci 47:330–350
Wickham H (2014) Tidy data. J Stat Softw 59(1):1–23
Youyou W, Kosinski M, Stillwell D (2015) Computer-based personality judgments are more accurate than those made by humans. PNAS 112:1036–1040

The manufacturer's authorised representative in the EU is Springer Nature Customer Service Centre GmbH, Europaplatz 3, 69115 Heidelberg, Germany. If you have any concerns regarding our products, please contact ProductSafety@springernature.com

Printed and bound by CPI Group (UK) Ltd, Croydon, CR0 4YY
30/09/2026
02225467-0007